MAKING BUILDINGS SAFER FOR PEOPLE

During Hurricanes, Earthquakes, and Fires

Contributors

Paul R. De Cicco, a professor emeritus at Polytechnic University, is one of the United States top experts on fire in buildings. He has extensive experience in full-scale fire testing and evaluating fire damage to building structures.

Ted V. Galambos, professor of civil engineering at the University of Minnesota, is a leading expert on steel structure design. He has played an important role in the development of several new design codes, and has authored or coauthored several textbooks on steel design and other topics.

Robert C. Murray, a member of the staff at Lawrence Livermore National Laboratory, is a leading authority on the evaluation of nonstructural damage caused by earthquakes.

Andrzej S. Nowak, associate professor of civil engineering at the University of Michigan, is a leader in the area of structural reliability and the modeling of human errors in structural design and construction. He has published more than 80 technical papers.

M. K. Ravindra, vice president EQE Engineering Incorporated (Costa Mesa, CA), is one of America's leading experts in the safety of nuclear plants, especially concerning seismic loads. He holds a Ph.D. from the University of Waterloo.

Norris Stubbs, a professor in the Department of Civil Engineering and in the College of Architecture at Texas A&M University, is an authority on the risk analysis of structures in a wind hazard. Dr. Stubbs, who holds a doctorate from Columbia University, New York, also has extensive experience in nondestructive evaluation of structures and the evaluation of quality in the design–construction process.

Preface

Safety analysis is becoming increasingly important. There is a need for more accurate and efficient methods to evaluate the adequacy of newly designed or existing structures. However, safety analysis should apply to both the structure (structural behavior) and the occupants (safety of occupants). This book deals with the latter issue.

The safety of building occupants is an important problem that affects millions of people. The consequences of fire, hurricanes, or earthquake can be greatly reduced, or even eliminated, if proper consideration is given to the safety of occupants during design and construction. A considerable amount of work has been done in the area of structural safety. There are methods available for the calculation of the probability of failure. The objective of this book is to review approaches to safety from the occupant's point of view.

The current design codes were developed by considering the structural behavior. The structural safety provisions are well addressed. For example, most of the newly developed design codes are based on probabilistic methods (American National Standards Institute, 1982; Ontario Ministry of Transportation and Communications, 1983; American Concrete Institute, 1986; American Institute of Steel Construction, 1986). However, there is a need to give more consideration to the safety of occupants in buildings. The importance of the safety of people is obvious when the structural failure records are considered. Recent building accidents clearly indicate that the legal consequences of human death or injury may considerably exceed the cost of structural damage (a classical example is the collapse of a walkway in the Hyatt Regency Hotel in Kansas City).

An examination of past failures indicates that, in many cases, some consideration given to the safety of occupants during the design and construction stage could result in a drastic reduction of the consequences of structural failure, in particular in the loss of life and limb. Therefore, there is a need for the inclusion of human safety aspects in the design and evaluation criteria. Considering the skyrocketing costs of insurance premiums, this can return large savings in this area.

The book may serve as a reference material for structural engineers, architects, and others involved in the building process. The risks associated with various categories of hazard are reviewed. Strategies for the reduction and/or elimination of risk are also discussed.

The objective of this book is to discuss the most important hazards affecting the safety of occupants in buildings, in particular, fires, earthquakes, and hurricanes. Issues addressed include the occurrences, mechanisms, and consequences of structural failure, and strategies to reduce the consequences thereof.

The text is divided into eight chapters. In Chapter 1 various categories of hazard are discussed, including the causes of structural and nonstructural damage, the consequences, and strategies to control them. The parameters of load and resistance are random variables. Therefore, probabilistic methods can be efficiently used to evaluate the safety of occupants. Safety measures and acceptable safety levels (risks) are discussed in Chapter 2. Procedures are also presented for the development of probability-based design criteria. The last six chapters deal with specific hazards. The discussion covers forms of occurrence, structural and nonstructural effects, characteristic cases, and strategies to reduce the consequences for occupants. Chapters 3 and 4 deal with the hurricane hazard, which is a major concern in many coastal areas of the United States. Chapters 5, 6, and 7 deal with the fire hazard, which is the dominating cause of building failures. Chapter 8 deals with the seismic hazard, a very important consideration for inhabitants of areas with higher seismic activity.

These chapters are written by experts with years of experience in dealing with hazards affecting building occupants. Practical examples are used to illustrate the discussed problems.

The original idea for the book was conceived during the ASCE (American Society of Civil Engineers) Structures Congress in Chicago in 1985, at the session organized by the editors of this volume. The chapters are written by the participants and speakers at the session.

The text is addressed to engineers involved in the design and development of design codes, to focus their attention on the important issues related to the safety of occupants.

Andrzej S. Nowak
Ted V. Galambos

REFERENCES

American Concrete Institute, 1986. Analysis and design of reinforced concrete guideway structures. ACI 358, *J. of the American Concrete Institute,* V. 85, no. 5, pp. 838–868.

American Institute of Steel Construction, 1986. *Load and Resistance Factor Design,* 1st ed. Chicago: American Institute of Steel Construction.

American National Standards Institute, 1982. *Minimum Design Loads for Buildings and Other Structures,* ANSI A58. 1–1982. New York: American National Standards Institute, Inc.

Ontario Ministry of Transportation and Communications, 1983. *Ontario Highway Bridge Design Code.* Downsview, ONT: Ontario Ministry of Transportation and Communications.

Contents

Chapter 1
Natural and Man-Made Hazards

T.V. Galambos
University of Minnesota

1.1 INTRODUCTION

This is a book on the safety of building occupants. The emphasis is on the occupants, and this chapter will elaborate on the risks they incur as they live, work, or play in and near buildings. This chapter will start by defining what a building is, what it is used for, and what the users and occupants expect of it. The hazards due to the occupancy and use of a building will then be discussed in more detail.

A building is a man-made structure that is erected according to a plan in order to serve a useful purpose. While some buildings are erected for monumental or ceremonial purposes, to demonstrate the omnipotence of God, the power of a famous leader, or the greatness of a civilization, most buildings serve as shelters, so that various human activities can be performed in comfort and privacy. A building protects its occupants from the environment. It keeps out noise, heat, cold, rain, and ice and the snow; generally its internal climate is controlled, and it protects the various umbilical cords that provide air-flow, water, electricity, telecommunication, waste-removal, and so forth. It also contains schemes for internal communication, such as hallways, stairways, and elevators; its individual cells are isolated from noise and vibration due to activities in other parts of the building. Most buildings also house the machinery needed to heat, cool, and ventilate the climate within it.

Buildings are used as places for living and working, they are used for gatherings, for sports activities, for religious worship, and for healing. Building sizes, shapes, and uses have an infinite variety, and modern large buildings are complex systems indeed.

Regardless of the great diversity of buildings, the one unvarying expectation of their occupants is that they be strong enough to support the expected loads throughout their intended life. This expectation is so pervasive that, for most buildings, not even the

1

suspicion of weakness (in the form of a perceived deflection or vibration) is acceptable. These attributes of strength and stiffness are provided by the *structure* of the building. This chapter will mainly concern itself with hazards associated with this structure.

1.2 A GENERAL DISCUSSION OF HAZARDS IN BUILDINGS

A hazard is the possibility of an event that could result in the death or injury of persons in the building, or in damage to property in the building, or in damage to the building itself. One way of dealing with hazards is for the occupants to be aware of all dangers, and to be constantly on the alert, so that when hazards arise they can be averted. This requires highly trained occupants who are ever-prepared for any eventuality. The other possibility is to construct a building without hazards, or at least to have a building containing machinery to automatically respond when hazards arise. Realistically speaking, it cannot be expected that either of these extreme situations are generally applicable; the actual cases are somewhere in between.

Hazards in buildings may be due to voluntary or involuntary actions by the occupants, they may be due to poor design or construction, or they may be due to the forces of nature.

There are many kinds of hazards in a building, some of which will not be discussed in detail, but which deserve a brief mention in passing, and some of which (such as the fire hazard) will be dealt with in later portions of this book. This chapter will deal with the hazards associated with the building itself. Some of the general hazards can be avoided by good common sense. Certainly, one should not smoke in bed, jump out of a window, light a match when one smells gas, use an electric device in the bathtub, or do any number of stupid things like that. Nevertheless, the majority of mishaps to occupants are such careless home accidents. Professor Lind, in the book *Methods of Structural Safety* (Madsen et al., 1986), quotes a 1977 study that concludes that a person's annual risk of death due to a home accident is about one in ten thousand. In contrast, a person's annual risk of death due to the structural failure of a building is about 1.0×10^{-7} (one in ten million). The risk of injury in a structural failure is about six times greater than the chance of death from such a failure. Comparatively, the annual risk of death per exposed person for automobile travel, construction workers, and work in a factory is 2.2×10^{-4}, 1.7×10^{-4}, and 0.4×10^{-4}, respectively. To quote Professor Lind: ". . .the risk of death from structural failure is very small. It is comparable to the risk of death from lightning and snake-bite. . . ."

It can therefore be said that the hazards to occupants in buildings due to structural failure are minuscule in comparison to other hazards. However, it must be emphasized that the risk is not zero. Furthermore, most accidents to which people are prone are the results of voluntary acts, while the hazards due to structural malfunction are involuntary. Occupants of buildings simply do not go to bed at night consciously worrying about the roof falling on them, while every reasonable vehicle driver is always on the lookout for situations that could cause an accident. Thus the risk of a structural failure in a building, while small, cannot be ignored.

The psychological make-up of people will not tolerate hazards from a structure that is provided to protect them from the awesome forces of nature.

1.3 PERCEPTIONS OF HAZARDS TO OCCUPANTS DUE TO STRUCTURAL FAILURE OF BUILDINGS

Structural design engineers plan and design the structure that provides buildings with adequate strength and stiffness. In creating the structure, they endeavor to optimize the demands of both structural safety and economy of construction. In their field, they use a mixture of science (e.g., structural mechanics, materials science) and judgment based on experience. They are guided in their design by the rules in building codes, which in turn refer to structural design specifications, material specifications, load codes, and so forth, which define the limits of safe structural design. These codes are constantly evolving, but at any point in time they represent the current composite judgment of various building professionals in the geographic area covered by the code as to what is a safe structure.

The judgments of structural engineering professionals and the concepts of the occupants in a building, as to what a safe structure is, may not be the same. The structural engineer is primarily concerned about the safety of the occupants, but is also concerned about the integrity of the structure, while the concern of the building occupants is more personal. The structural engineer thinks of failure in terms of exceeding a mathematically defined limit state: the yielding of a member, buckling of a plate, instability of a frame, failure in a connection, etc. The occupant is worried about the well-being of his own person and of his family or associates. The occupants' worries are about more tangible manifestations of malfunction: the total collapse of a building, cracking in a member, sagging of a floor, vibration of the structure, etc. In general, structural engineers have a more strict interpretation of the risk of failure than the occupants. For example, occupants may be totally unaware that their building is unsafe in a severe earthquake, simply because such an earthquake has never occurred. On the other hand, the engineer, upon examination and analysis, may condemn this building. There are, however, exceptions to the general rule that engineers are more strict than occupants in interpreting structural risk: occupants may be thoroughly frightened by the swaying of a chandelier in a restaurant on top of a tall building during a thunderstorm, while the science and experience of the structural engineers tells them that this swaying is perfectly natural and that it poses absolutely no harm to either the building or the patron of the restaurant. Nevertheless, the building has failed in the mind of the occupant, and he or she manifests this conviction by never returning to this particular establishment.

Structural engineers like to think of a building as consisting of two types of systems: the *structural system* (i.e., the beams, girders, columns, slabs, trusses, shells) and the *nonstructural* system (i.e., curtain walls, partitions, etc.). To the lay occupants, such distinctions do not exist, and they consider the structure to be damaged regardless of the type of element involved. These occupants are probably more correct than the engineers. Damage to a nonstructural component often indicates trouble with the structure itself. In any event, both types of element interact, under severe loading, in more complicated ways than are possible to idealize for structural analysis.

1.4 CATEGORIES OF HAZARDS DUE TO STRUCTURAL FAILURE

We will distinguish two hazard classifications: 1) man-made hazards and 2) natural hazards. Either of these can have varying degrees of severity, listed from most to least serious (from the point of view of structural engineers): 1) life-threatening hazards, 2) hazards of economic loss, and 3) nuisance hazards (noise, vibration, etc.).

1.4.1 Man-Made Hazards

These hazards are due to some human action, which are mostly inadvertent, but which could also be willful.

Hazards Resulting from Poor Structural Design. Individuals or groups of people cannot do anything without the possibility that mistakes will occur. This fact is recognized in the engineering design process, and designs are checked and verified in various ways by the organization creating the design, and by external control agencies. The majority of design errors are rectified before or during construction. Furthermore, conservative design assumptions and factors of safety, while not intended to cover errors, nevertheless tend to protect the occupants from most of the remaining errors. However, errors can slip through and may eventually pose a hazard.

A list of the most prevalent errors in design are discussed in the following paragraphs.

Ignoring a load type that could possibly occur during the life of a structure. A prevalent example of this hazard is in ignoring earthquake loads in areas of dormant seismicity. There are virtually millions of buildings in the United States, particularly in the Mississippi basin, that could pose significant life-threatening hazards to the population if earthquakes, which occurred near New Madrid, Missouri, in 1812 and 1813, were to happen again. Another frequently found hazard is in neglecting to consider the possibility of snow drifts on roofs. One additional hazard in this category is in not accounting for the possibility of roof failure from the ponding of water or ice.

Errors of calculation in the structural design process. Such errors are usually caught, or they may be inconsequential. Yet, there is a finite, if small, chance of hazard from this cause.

Errors in judgment. Structural engineers, in order to be able to calculate forces, must idealize the structure into an abstraction. The process of modeling the structure is, next to the modeling of the loads, the most important and creative aspect of the structural engineer's job. While the modeling is more-or-less routine for simple structures, it takes a great deal of experience and understanding to abstract a complex structure into a model that faithfully mirrors reality. The most prevalent errors in judgment concern the inadequate provision for lateral bracing of members (this error caused the collapse of the Hartford Arena) or for the entire frame. Other errors caused by a lack of proper judgment occur at the very early stages of planning, where decisions are made about the actual form of the structure. It is important to design a structure that is not overly vulnerable, but is robust enough to resist very unlikely, but still possible, events. A structure should not entirely collapse if subject to, say, a gas explosion in a kitchen in an apartment house, or an impact by a runaway truck on an exposed corner column in a factory. In earthquake areas, care must be exercised not to design structures with open floors or with sharp changes in geometry, so as not to attract large forces to locations of weakness or stress concentration. Large eccentricities between the center of mass and the center of stiffness should be avoided. In these days of widespread terrorism, it is also prudent, when designing important governmental buildings or lifeline structures such as hospitals, communications centers, or power plants, to provide a structure that is not vulnerable to the effects of bombs or missiles.

One of the attributes of a good building is that its structural scheme should possess sufficient redundancy so that it is robust enough not to be subject to progressive collapse due to totally unexpected, but still possible, violent causes. Back-up systems or alternate load paths should be routinely designed into structures.

Neglect of dynamic load effects. Structural designers, in common with the lay occupants, often consider the building to be entirely static in its response. Occupants are particularly sensitive to motion of even the slightest magnitude, particularly in buildings where the activities are more-or-less sedentary, such as in libraries, churches, and schools. Structural designers are, of course, very aware of dynamic effects that could have life-threatening consequences, such as severe storms or large earthquakes, and these effects are accounted for in the design. They are generally less aware of smaller motions manifesting themselves as steady or transient vibrations. These vibrations could be due to rotary machinery in or near a building, traffic from vehicles in a building, from gusty winds buffeting the outside of the structure, or from such routine acts as walking, jumping, running, or dancing. The designer has at his disposal a vast literature on how to avoid these small vibrations, and prudent engineers know how to avoid them. Nevertheless, they are often ignored in design because they pose no threat to safety, the design calculations are often quite elaborate, and the costs of dealing with them may be quite high. However, occupants are usually frightened by such vibrations, and they are generally perceived to be hazardous. These vibrations are, therefore, a very great nuisance, and their reduction (to levels that are not annoying) is usually very expensive after the building is completed.

The hazards to the occupants from poor structural design can be avoided by wise and careful design, and their occurrence is quite rare.

Hazards Resulting from Poor Construction.   The process of construction gives ample opportunity for the creation of hazards to the building occupants. This fact is also recognized in practice, and there are many levels of control and inspection throughout the construction activity to alleviate them. Large mistakes during construction usually result in failure while erection is still going on, and so the remedies are applied before the occupants move in. Despite this self-correcting situation, and despite the many controls, it is still possible to introduce faults during construction that could result in eventual dangers to the inhabitants. An enumeration of such hazards is contained in the following paragraphs.

Incorrect construction practices.   In spite of the usual great care exercised by construction workers and the supervisors who inspect their work, it is still possible that minor faults are introduced into the structure, which could result in hazards later, when the building is subjected to a major storm or earthquake. In a large building, there may be literally hundreds of thousands of reinforcing bars, or structural welds or bolts. It is impossible that all of these elements are free of defects, or that all of these are detected or repaired during construction. Experience has shown that, under usual construction practice, the hazards from such errors (due to faulty materials and incorrectly placed elements) are very small. However, shoddy construction or erection under unduly severe time and fiscal constraints will produce more hazards. Many of these will eventually become known during occupancy, and their alleviation represents a costly inconvenience to the owners of such buildings.

Inadequate communication. Modern construction practice is a very complex operation. The evolution of building construction in the United States has resulted in the situation that each new building is created by a set of participants that join together for the express purpose of erecting this particular artifice. Upon the completion of this work, this composite organization is dissolved, to form again into a different grouping with different actors for another construction project. The creators of the building consist of the construction manager, the financial backers, the owners, the architects, the engineers (e.g., foundation, structural, materials, electrical, and HVAC specialists), the general contractor, and the subcontractors (e.g., the fabricators, the erectors, the curtain-wall manufacturers, the electrical contractors, and the HVAC installers), to mention some of the participants. In addition, the workers represent a variety of trade unions, each operating under their own rules. Finally, there may be a diversity of requirements, supervision, and controls from government agencies having jurisdiction over the building. Indeed, the creation of a new building is an awesome and complicated art. The key to success is in communication between all of the actors, aided by good planning and by an efficient organization of tasks. This system of construction, despite its complexity, works remarkable well. However, things can go wrong, as attested to by the Hyatt Regency disaster in Kansas city in 1981. In that case, an incorrect change to a detail was either not communicated to the designer, or it was not properly checked out. While the disaster was a fatal tragedy to the 120 or so fatalities and their families, some good came of it in that subsequent professional and legal actions resulted in the clarification and formalization of the lines of responsibility in construction.

Hazards Resulting from Human Action During Occupancy. While the chance of creating hazards to the eventual users is high during design and construction, there are many controls and remedial measures which reduce them to the point that, when the building is finally ready for occupancy, the occupants can expect a very high degree of reliability. There are few such controls during the occupancy phase, where avoidance of hazards depends essentially on the prudence, common sense, and good will of the occupants themselves. The following actions can result in hazards while the building is occupied.

Deliberate acts of sabotage, terrorism, or vandalism. Even though the designers deliberately, or from intuitive good sense, plan a robust structure, there is no real defense against such acts when they are perpetrated by determined criminals.

Overloading. Hazards to occupants can result from inadvertent or careless overloading of some parts of the structure. In the design process, the load factors applied to the expected loads foresee some overloading (e.g., the design live loads contain an allowance for the crowding of people or piling up of furniture), but there is no way to account for excessive misuse. Occupants should not radically increase the loads before assuring themselves that the structure will not be damaged. This is particularly so when installing items such as heavy machinery or steel vaults.

Modification of the structure. Occupants are usually intelligent enough not to tamper with beams or columns; it is possible, however, that they could remove walls that are either load-bearing or serve as part of the lateral load-resisting or stabilizing system of the structure. As with possible overloading, a structural investigation should precede any substantive modification of the building.

Explosions. Certain industrial processes in a plant using explosive chemicals or the explosion of cooking gas, etc., can present a hazard to the structure. Such hazards should be considered in building design, and proper care should be exercised by the people using the building. However, explosions are still a possible hazard.

Fire hazard. This danger will be examined in greater detail later in this book. Fire not only endangers the lives and property of the occupants, but it can also represent a hazard to the structure itself, if its elements are not provided with adequate fire-proofing. The code requirements relating to fire-proofing are stringent, and they are strictly enforced, so this particular hazard is a relatively small danger.

Corrosion. Some industrial processes, or excessive dampness (say, in a basement), can result in corrosion of structural elements. This can result in loss of cross section and in the eventual failure of a member of the structure.

Collision. Some buildings are subject to impact from vehicles operating in and around them. Impact from airplanes is an ever-present hazard near airports.

There are many human actions, either deliberate or inadvertent, that could damage a structure, and thus pose hazards to the people occupying a building. Fortunately, the structure is usually robust enough so that the actions and forces capable of being generated by humans without construction machinery pose a negligible threat.

1.4.2 Natural Hazards

Natural hazards are accounted for in building design in the form of forces, accelerations, and impacts which the structure must be able to safely resist. These design actions are generally prescribed in great detail in the codes and regulations. In unusual cases, say for tall buildings of unusual shape, wind forces are often determined from wind tunnel studies. The design of structures is based on the most extreme natural forces expected during their intended life. These design forces are then increased by load factors that attempt to account for the uncertainties of the load determination. Both the structure and its occupants are, therefore, very well protected from the consequences of natural hazards. Unfortunately, it is still possible that natural hazards pose dangers to the occupants. There are several reasons for this. Some of these have been mentioned previously, but they, and some new ones, are elaborated in the following.

Areas of Weakness Introduced by Poor Design or Poor Construction. It is possible, through errors of judgment or calculation, or mistakes or omissions in construction, to introduce flaws or regions of weakness that could affect the structure adversely in the event of an extreme natural catastrophe, such as a large earthquake or tornado. The literature on the investigation of damaged buildings after severe earthquakes, large snowstorms, and heavy winds abounds with examples where collapse or damage could be directly related to such flaws in the structure or in the design assumption.

Unusual Natural Occurrences. Our knowledge of the forces of nature is far from complete. This knowledge is based on the past history of catastrophes in the region, on our understanding of the laws of physics governing the natural phenomena of wind, snow, and earthquakes, and (in recent times) on records obtained from instruments placed on or near structures. As was so recently demonstrated in the Mexico City earthquake of September 1985, there are always new facts about nature that were not fully appreciated beforehand. The science of predicting the expected accelerations due to a severe earthquake at any given building site is just not complete enough to hope that all affected structures will ride out such a shake without damage. In the opinion of the author of this chapter, this is the most serious type of hazard that occupants of buildings will ever be exposed to. Severe natural events are so rare and so complex that the element of a novel twist is always present. Fortunately for us, these catastrophes are not only rare in time but also restricted as to the area affected. Furthermore, current research is very active in this field, and every year brings forth new knowledge that will eventually benefit the safety of us all.

Unexpected Natural Hazards.   Such hazards are related to the ones discussed above, but in this case the designers have concluded, in all good faith based on sound judgment and on no previous historical occurrence, that a particular natural phenomenon cannot be expected to ever affect the structure. Examples of this would be a severe snowstorm in Florida, a hurricane in Montana, or a volcanic eruption in Minnesota. While these examples are far-fetched, they illustrate the possibilities. Again, the most serious of such events are earthquakes. Had we known all that we know today, would we have designed structures in Charleston, S.C. against earthquakes prior to the earthquake of 1886?

Deliberately Ignored Natural Hazards.   For reasons of economy, most structures in the American Midwest are not designed to resist tornadoes, despite the frequent occurrence of these storms. This is a deliberate economic decision, based on the observation that tornadoes touch down and destroy relatively small areas, and thus the chance of any given building being hit by a tornado is so small a risk that widespread construction for tornado-resistance is not economically justifiable. Nevertheless, the risk is real, as witnessed by the annual toll of lost lives and destroyed property.

Water Hazards. Water can present light to serious hazards to buildings. Obviously, floods can damage or destroy buildings in areas affected by them. However, water can do other harm to structures: water may pond to such an extent, from rain on a large flexible roof, or from a break in a water pipe on a floor, that it brings down the structure. Water may also damage inadequate exterior wall systems so as to seriously impair their strength. Excessive moisture can cause severe corrosion of structural elements.

Wind Hazards. In addition to the wind effects discussed previously, strong winds can hurl missiles, which could place people and property into grave danger. Such missiles could also damage structures, and especially sensitive structures should be designed to resist these windborne objects.

Earthquake Hazards. In an earthquake-resistant structure, even in a moderate earthquake, occupants and their property can be damaged by objects that are accelerated by the sudden motion of the building. The televised pictures of the mess inside homes, offices, libraries, and grocery stores taken after the July 1986 earthquake in Palm Springs, California, illustrate this vividly. Simple precautions could almost entirely eliminate this hazard.

Foundation Failure. Floods and earthquakes can severely disrupt the foundation of buildings, thus causing structural distress and even failure. This hazard can be mostly eliminated by proper design of the foundation or by judicious site selection.

Aging. A final natural hazard deserves mention here. Materials and structures eventually grow old. They eventually deteriorate to the point where the structure is hazardous. This aging process is accelerated in moist and cold climates, particularly for structural elements that are only partially protected from heat, cold, and moisture.

1.5 CONSEQUENCES OF FAILURE

In the previous section of this chapter, a veritable litany of hazards to which occupants of buildings are exposed to was recited. Upon reflection, readers could add many more additional horror stories.

Despite the many things that can go wrong, very few hazards actually become active to the extent that they affect the safety and the pocketbook of the occupants.

The consequences of structural malfunction, damage, or partial or full collapse, when they occur, are usually very expensive. In the case of accidents resulting in death or injury to the occupants, the cost cannot be really enumerated in financial terms. Economic loss incurred in repairing damage can be very expensive, and this cost vastly overshadows the initial cost of doing the job correctly in the first place. The following chapter will further elaborate on the economic consequences.

The possible consequences of structural failure are briefly listed as follows:

1. Loss of life.
2. Impairment of the physical and mental health of the occupants.
3. Economic loss to the designers, builders, occupants, and owners.
4. Loss of confidence in the professionalism of design professionals or contractors.
5. Loss of the use of a facility important to the community.
6. Economic loss to an entire segment of the building industry.

1.6   ACTIONS AVAILABLE TO AVOID HAZARDS IN BUILDINGS

The major actions to limit hazards to the occupants of buildings must be taken by the designers and planners responsible for their construction. These professionals are legally required to have, as their primary agenda, the physical and fiscal well-being of the people who will use the buildings throughout its entire life-cycle. This responsibility is emphasized daily in the education of these professionals; in building codes, which define the limits of design criteria; and in extensive and constant internal and external checking and control measures, to assure the elimination of most errors. There is thus in place a system of professional procedures that is both conscious of possible hazards, and which, theoretically and practically, minimizes them. Furthermore, design professionals have a high set of ethical standards that guide their actions.

As seen in previous parts of this chapter, hazards can still exist, despite all of the precautions and despite all of the care, simply because human activities are never error-free, and because mankind is never completely in charge of its fate. We can, however, do our best, and history shows that this is not inconsiderable. We could, however, improve performance through the use of several measures that are not now necessarily part of the design process:

1. Designers and planners should be given more time and more financial resources to completely consider all eventualities.
2. Building codes and design standards should reflect more awareness of hazards, and give more explicit aids to designers.
3. More time and resources should be made available for the supervision of construction.
4. All actors in the design and construction processes of buildings should get together in meetings to determine issues of constructability and to work through hazard scenarios, so that all affected parties would become aware of the possible eventualities of malfunctions. This kind of an exercise would especially benefit the construction of particularly complex buildings, and buildings that must remain functional during and after a severe natural catastrophe or a terrorist attack.

Designers, planners, and constructors should thus continue what they are doing now, improving their performance as time and fiscal constraints permit, to provide buildings that are as free of hazards as possible. Occupants, on the other hand, should also do their share in this process. They can easily do the obvious things:

1. Make no modification to structures without consulting building officials or engineers.
2. Take care to not overload or otherwise misuse the structure.
3. Participate in no hazardous acts that could result in damage to the structure.
4. Report any changes in structural performance (e.g., vibrations, deflections, cracks, corrosion, moisture, water or ice-ponding) to those responsible for the building.
5. See to it that the building is properly inspected and examined after a major natural catastrophe or any violent human act.

6. Realize that (despite all care and caution exercised in design and construction) the building is a human artifact, which may contain hidden flaws, and that no system of construction, be that man-made or natural, is absolutely perfect and without hazard.
7. Take common sense actions to prevent injury or economic loss due to motion of a building (i.e., make sure that furniture cannot slide or fall over in an earthquake).

1.7   SUMMARY AND CONCLUSION

This chapter has attempted to show that buildings, as presently used and constructed, are indeed very safe, and that occupants have no reason to be unduly apprehensive about their well-being while living, working, or playing in a building. An attempt has also been made to show that no building system is completely free of hazards. Some of these hazards can be repaired once they are detected, and some are small enough to only be a nuisance. However, there can be undetected hazards that can lie dormant, but which can act as a fuse, causing damage or even collapse during an extreme natural or terrorist event. Hazards can be minimized by proper design and construction and by the vigilant and prudent behavior of the occupants. Finally, no one should expect that error can be completely eliminated, or that the forces of nature cannot play tricks that are entirely unanticipated.

REFERENCE

Madsen, H. O., Krenk, S., and Lind, N. C., 1986. *Methods of Structural Safety*, 403 pp. Englewood Cliffs, NJ: Prentice-Hall, Inc.

Chapter 2
Risk Analysis

Andrzej S. Nowak
University of Michigan, Ann Arbor, MI

2.1 INTRODUCTION

Risk represents the possibility of loss or injury. Any human activity involves a certain exposure to risk. Absolute safety is impossible to provide. An indication of the degree of risk is the number of accidents, failures, injuries, or deaths incurred by participants in the activity. Intuitively, potentially dangerous courses of action can often be identified. However, quantification of risk is very difficult. There are many parameters involved, including cause (direct or indirect), items or persons affected, the number of persons, their age, social status, professional status, consequences (short-term and long-term), and costs. Risk analysis is an important tool in the selection of an appropriate safety level.

Some statistical data on death rates as a function of exposure to certain activities has been presented by Melchers (1987), as shown in Table 2.1. The numbers corresponding to the structural failure are at the bottom of Table 2.1. However, the observed values do not necessarily reflect the human perception of the risk.

In this chapter, safety levels used in current engineering practice are discussed and compared. Values are recommended for serviceability limit states concerning the safety of occupants in buildings.

2.2 UNCERTAINTIES IN BUILDING PRACTICE

The risk associated with the occupation of buildings is due to uncertainties in loads, user's actions, and the behavior of structural and nonstructural elements. These uncertainties, in turn, result from natural variations in loads and material properties, man-made hazards (gas explosion, fire, collision), insufficient knowledge of loads and structural behavior, and human errors in design, construction, and use. In the last twenty years, considerable developments have been made in the area of structural reliability (Thoft-Christensen

Table 2.1 Risks associated with Various activities (*from* Melchers, R. E. *Structural Reliability, Analysis and Prediction*. Reprinted with permission from Ellis Horwood Limited, Chichester, U. K. © 1987).

ACTIVITY	DEATH RATE (10^{-9} DEATHS/HR EXPOSURE)	EXPOSURE (HR/YEAR)	DEATH RISK (10^{-6}/YEAR)
Alpine climbing	30,000–40,000	50	1500–2000
Boating	1500	80	120
Swimming	3500	50	170
Smoking	2500	400	1000
Air travel	1200	20	24
Car travel	700	300	200
Train travel	80	200	15
Coal mining	210	1500	300
Construction work	70–200	2200	150–440
Manufacturing	20	2000	40
Building fires	1–3	8000	8–24
Structural failures	0.02	6000	0.1

and Baker, 1982; Madsen et al., 1986; Melchers, 1987). Methods are now available for the calculation of the probability of failure for given load and resistance models. New design codes are calibrated to provide a preselected target reliability level. It is now becoming possible to control the degree of risk.

The major means of risk control can be put into the following categories:

1. Control of causes
   - Passive control, by reduction of the exposure to causes, with regard to time, surface, presence, the number of people, the weather, environmental conditions, natural hazards, and man-made hazards.
   - Active control, by the elimination of sources, and by checks and inspections.
2. Control of consequences
   - Early warning systems, to allow time for preparation for unloading, change of occupancy, repair, strengthening, or evacuation.

- Failure isolation, to prevent progressive collapse and to allow for uninterrupted functioning of the undamaged parts of the building.
- The use of fail-safe systems. If failure is to occur, it should affect only less important and less consequential parts and/or limit states.

The selection of the control system is an economical problem. Safety can be considered as a commodity, which can be purchased for a current price. The optimum safety level corresponds to the minimum utility value. An important tool in this calculation is the quantification of safety or the probability of failure, which is the subject of the theory of reliability. A practical methodology is now available to evaluate the reliability of a structural member or structural system. However, little has been done to consider the safety of occupants in buildings.

2.3 RELIABILITY OF STRUCTURES IN CURRENT PRACTICE

The performance of structural members can be measured in terms of the probability of failure. A structure can be in a safe state (the load effect is less than load-carrying capacity), or at failure. It is convenient to express both the load effects and resistance in terms of parameters (load components, material properties, or dimensions). Then the state of structure can be described using a limit state function. In the simplest form, there are two random variables, R equalling resistance and Q equal to load effect. Then, the limit state function, g, is

$$g = R - Q \qquad 2.1$$

where a negative value of g indicates a failure.

An example of the probability density functions (PDF) for resistance, f_R, and load effect, f_Q, is shown in Figure 2.1. The resulting PDF for g, corresponding to f_g, is also shown, with the probability of failure, P_F, equal to the shaded area. The corresponding joint probability density function of R and Q is shown in Figure 2.2. The probability of failure is also equal to the volume obtained by the integration of this density function over the failure region (R less than Q).

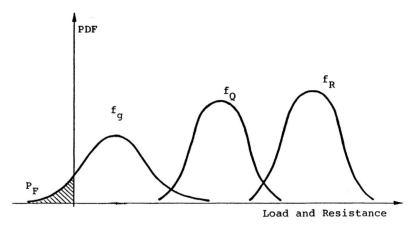

Figure 2.1. Density functions of load, resistance, and safety margin.

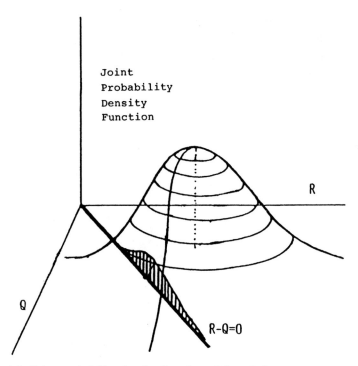

Figure 2.2. Joint probability density function of R and Q.

The probability of failure can be calculated using special formulas and numerical procedures. It is convenient to introduce a reliability index, B, as a measure of structural performance,

$$B = - F_N^{-1} (P_F) \qquad 2.2$$

where F_N equals the standard normal probability function. Some values of B, and corresponding P_F, are given in Table 2.2.

In general, the load effect is a joint effect of several components (the dead load, live load, environmental loads, and others). Also, the resistance depends on several parameters (the strength of material, geometry, degree of deterioration). Let X_1, . . . , X_n be random variables, representing the load and resistance components and other parameters. Then R and Q in Equation 2.1 can be replaced with the functions of X_i's, and the limit state function can be expressed in terms of X_i's too.

The limit state function and the associated probability of failure are clearly defined in the case of ultimate limit states, such as the bending moment capacity, or the ultimate shear capacity of single members. However, the serviceability limit states, such as cracking, fatigue, deflection, or vibration, often govern the design. These are also the major factors determining the safety of occupants in buildings. The limit state is often difficult to define. The main concern can be the accumulation of damage caused by repeated applications of load (fatigue or cracking). Therefore, reliability analysis requires

Table 2.2 Reliability index and probability of failure.

RELIABILITY INDEX	PROBABILITY OF FAILURE
0	0.5
1	0.159
2	0.0228
3	0.00135
4	0.0000317
5	0.000000287
6	0.000000000987
7	0.00000000000128
8	0.00000000000000000611

knowledge of not only load magnitude, but also frequency of occurrence.

Structural safety analysis methods have been developed in the last 20 years. For a formulated limit state function and parameters of load and resistance, there are various procedures available to calculate the reliability index (Thoft-Christensen and Baker, 1982 and Melchers, 1987). The methods vary, depending on the accuracy, computational effort, and required input data. For the ultimate limit states, typical values of B were calculated for buildings (Ellingwood et al. 1980) and for bridges (Lind and Nowak, 1978). Examples are shown in Table 2.3 for buildings, and in Table 2.4 for bridges.

For serviceability conditions, the limit state function needs special formulation. In the case of fatigue, R and Q can be expressed in terms of the number of load cycles (R equalling the number of cycles to failure, and Q equal to the number of cycles applied). For a given load level, the number of cycles to failure, N_F, is a random variable, with a PDF of f_{NF}. The number of cycles applied, $N(t)$, can be also treated as a random variable, a function of time, t, with a PDF of $f_{N(t)}$. The relationship between these variables is shown in Figure 2.3. If, however, the load magnitude varies, then the formulas become more complex (a well-known Miner's rule can be used). An example of such a formulation for prestressed concrete girders was presented by Al-Zaid and Nowak (1988).

Table 2.3 Reliability indices for buildings (*from* Ellingwood et al., 1980).

TYPE OF MATERIAL	LIMIT STATE	RELIABILITY INDEX	PROBABILITY OF FAILURE
Cold-formed steel		2.5	6.2×10^{-3}
Hot-rolled steel	Tension	3.4	3.4×10^{-4}
	Compression	2.2–3.1	
Reinforced concrete	Flexure	2.7–3.5	$1.4 \times 10^{-2} - 9.5 \times 10^{-4}$
	Compression	2.5–4.5	$6.2 \times 10^{-3} - 3.4 \times 10^{-6}$
	Shear	2.5	6.2×10^{-3}
Wood	Flexure and tension	2.0–2.5	$2.3 \times 10^{-2} - 6.2 \times 10^{-3}$

Table 2.4 Reliability indices for bridges (Lind and Nowak, 1978).

TYPE OF MATERIAL	LIMIT STATE	RELIABILITY INDEX	PROBABILITY OF FAILURE
Steel girders	Flexure	4.5	3×10^{-6}
	Shear	6.5	4×10^{-11}
Pretensioned concrete	Flexure	6.0	9×10^{-11}
girders	Shear	3.5	2×10^{-4}
Post-tensioned concrete	Flexure	6.0	1×10^{-9}
decks	Shear	4.0	3×10^{-5}

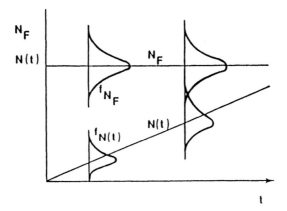

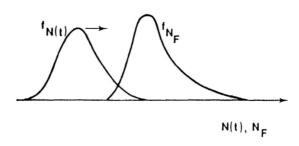

Figure 2.3. Fatigue limit state. The number of load cycles to failure, N_F, versus the number of cycles applied, $N(t)$.

2.4   ACTUAL AND PERCEIVED SAFETY LEVELS

The safety of structures, or the probability of failure, is definitely not uniform for all structural types, materials, spans, uses, etc. Reliability indices calculated for structures designed using current code provisions show considerable scatter. For example, reliability indices were calculated for bridges in Ontario designed using old code provisions. For comparison, the calculations were also done for bridges designed using calibrated probability-based load and resistance factors. The resulting values of B_{old} and B_{new} are shown in Figure 2.4 for three types of bridges: composite steel girders, pretensioned concrete girders, and post-tensioned concrete decks (Lind and Nowak, 1978). Many code-developing committees are presently working on the rationalization of design procedures on the basis of probabilistic methods. One of the objectives is to reduce the scatter of B values. However, it is not necessarily desirable to obtain a uniform safety level for all structures and designs.

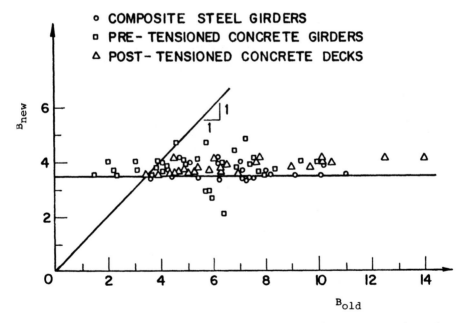

Figure 2.4. Reliability indices for bridges prior to calibration, B_{old}, and after calibration, B_{new} (*from* Lind and Nowak, 1978).

The required safety level is established in a complicated, and not always clear, procedure. The major document specifying the minimum acceptable reliability is the building code. However, there is usually no direct reference to reliability in the code. Instead, it is done indirectly, by specifying the minimum required section of the member or strength of material. The code provisions can be changed. The code-writing committee may adjust the reliability level of the designed structures. If there are problems with existing or newly constructed structures, then an increased target reliability level is needed, and if the present practice seem to be overly conservative, then a reduced target reliability level can be allowed. Evaluation of the reliability may serve as a rational basis for this decision-making process.

The human perception of risk is not always rational. Spectacular failures, well publicized by the media, often attract an unproportional share of attention and funds. Examples can be the nuclear industry (an irrational increase of anxiety with regard to reliability after the Chernobyl disaster), or the aircraft industry (increased public interest and requirements for a higher safety reserve immediately after airliner crash). People often do not realize the degree of risk involved in various situations. Society seems to pay much more attention (and money) to the prevention of natural disasters immediately after the occurrence of a major earthquake, flood, or fire, as compared to before such occurrences. This is reflected in fluctuating changes in design codes, the designation of research funds, and public awareness in general. It is the obligation of the engineering profession to inform society about the actual risks, so that the available limited resources can be distributed rationally and most efficiently.

Typically, the public does not expect any failures of building structures. This can be seen from the reaction to failure reports in the media. On the other hand, the observed rates of failure are higher than theoretical values (Brown, 1979). Some actual rates are shown in Table 2.5. For comparison the probabilities of failure due to man-made hazards are shown in Table 2.6 (Leyendecker and Burnett, 1976). The current situation can be considered to be a certain status quo. Even though expectations are high, limited funds are spent on failure prevention causes (additional checking and inspection, larger safety reserves, more research for the development

Table 2.5 Real-life rates of failures.

TYPE OF STRUCTURE	RATE OF FAILURE
Long-span suspension bridges	$1:40$
Large cantilever bridges	$1:70$
Tall buildings in the U.S.A.	$1:700-1:800$
Earth dams, up to 465 ft.	$1:25$
Bridges in the U.S.A. (per annum)	$1:3500-1:4000$

of more reliable structures). Practically, present failure rates are acceptable. Engineering practices change gradually over the years. The process is rather slow, and the safety reserves (structural reliability) stay at a constant level for many years, which can be treated as an indication that society accepts the present practice.

2.5 ECONOMIC ANALYSIS

Optimum rates of failure (structural reliability) can be derived by considering the costs. The total cost, C_T, is a sum of the initial cost of design and construction, C_I, and the expected cost of failure,

$$C_T = C_I + P_F C_F \qquad 2.3$$

where P_F equals the probability of failure, and C_F equals the cost of failure.

The probability of failure can be reduced, but this may cause an increase in the initial costs (higher safety factors, more checking and

Table 2.6 Probabilities of failure due to man-made hazards (*from* Leyendecker and Burnett, 1976).

TYPE OF ABNORMAL LOADING	PROBABILITY OF OCCURRENCE PER ANNUM PER DWELLING UNIT		
	MINOR CONSEQUENCES	INTERMEDIATE CONSEQUENCES	SERIOUS CONSEQUENCES
Gas explosion	18×10^{-6}	2.5×10^{-6}	1.6×10^{-6}
Vehicle collison	600×10^{-6}	86×10^{-6}	7.8×10^{-6}
Bomb explosion	0.92×10^{-6}	0.12×10^{-6}	0.22×10^{-6}
Airplane collision		8.2×10^{-6}	

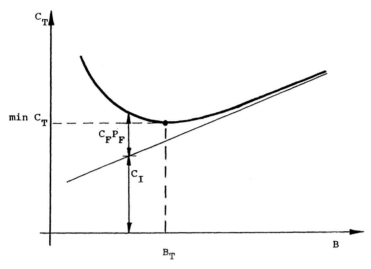

Figure 2.5. Relationship between the total cost and the reliability index.

inspection). The cost of failure, C_F, usually does not depend on C_I. The optimum P_F minimizes C_T. Mathematically, this corresponds to zero of the derivative of C_T, with regard to P_F. An economic analysis was performed by Lind (1976) for steel structures. He observed that C_I is a linear function of B (the reliability index) for the practical range of values. The resulting relationship between B and C_T is shown in Figure 2.5. The optimum value of the reliability index, B_T, corresponds to the minimum total cost.

In practice, an additional review or checking procedure may have a considerable effect on the reliability, with a negligible increase of C_I. The relationship between the reliability and checking intensity requires further study.

2.6 PROPOSED APPROACH

There is a need for rational acceptance criteria for various conditions affecting the safety of occupants in buildings. The development work may include the following steps:

1. Identifying the conditions to be considered (earthquake damage, hurricane or heavy wind pressure and effects, fire, gas explosions, and so on).

2. Formulating the limit state function(s) corresponding to these conditions, and identifying the parameters and gathering statistical data (distributions, correlations, time-variation).
3. Evaluation of safety level(s) for the considered conditions (limit states), corresponding to current design and construction practices. A spectrum of reliability indices will be obtained.
4. Evaluation of the current practice, using the reliability index spectra, expert opinions (Is the current practice acceptable?), and economic analysis, and establishing target reliability level(s) for the considered conditions (limit states).
5. Development of design and evaluation criteria corresponding to the established target reliability indices, and the calibration of load and resistance factors, and specifying special control procedures (checking and inspection) providing for these safety levels.

REFERENCES

Al-Zaid, R. and Nowak, A. S., 1988. Fatigue strength of prestressed concrete girder bridges. *Canadian J. of Civil Engng.*, Vol. 15, no. 2, April 1988, pp. 199–205.

Brown, C., 1979. A fuzzy safety measure. *J. of Engng. Mechanics Div., ASCE,* Vol. 105, no. EM5, Oct. 1979, pp. 855–872.

Ellingwood, B. et al., 1980. *Development of a Probability Based load Criterion for American National Standard A58.* Washington, D.C.: National Bureau of Standards Publication 577, June 1980.

Leyendecker, E. V. and Burnett, E. F. P., 1976. *The Incidence of Abnormal Loading in Residential Building.* Washington, D.C.: National Bureau of Standards Science Series no. 89, Dec. 1976.

Lind, N. C., 1976. Approximate analysis and economics of structures. *J. of the Structural Div., ASCE,* Vol. 102, no. ST6, June 1976, pp. 1177–1196.

Lind, N. C. and Nowak, A. S., 1978. *Calculation of Load and Preformance Factors.* Report, May 1978. Department of Civil Engineering, University of Waterloo, Ontario.

Madsen, H. O., Krenk, S., and Lind, N. C., 1986. *Methods of Structural Safety.* Englewood Cliffs, NJ: Prentice-Hall, Inc.

Melchers, R. E., 1987. *Structural Reliability, Analysis and Prediction.* Chichester, West Sussex, U.K.: Ellis Horwood Limited.

Thoft-Christensen, P. and Baker, M. J., 1982. *Structural Reliability Theory and Its Applications.* New York: Springer-Verlag.

Chapter 3
Hurricane Hazard: A Theory of Occupant Safety

Norris Stubbs
Texas A & M University, College Station, TX

3.1 INTRODUCTION

Undoubtedly, coastal areas are pleasant places to live and work, and indeed they should be conscientiously developed for such purposes. However, such areas (e.g., the Gulf and Atlantic coasts of the United States) are constantly threatened by hurricanes. Hurricanes can impart significant undesirable effects to the environment and society. When they strike, hurricanes can decimate the man-made environment. Buildings and roads can be heavily damaged or destroyed, public utilities can become dysfunctional, and the normal flow of business can cease. Hurricanes can also significantly modify the natural environment. For example, large stretches of sandy beaches may disappear, significant offshore reserves of food resources may be destroyed, and the equilibrium of the coastal ecosystem may be severely disrupted. The hurricane's impact on society may also be devastating. The collective stress borne by a community in coping with events like hurricane warnings, evacuation recommendations, sheltering decisions, anticipation of losses to personal property, the perceived risk of injury or death, and the inconvenience of the recovery phase—in short, the total disruption of ordinary life—is enormous. In addition, the economic costs resulting from the destruction and disruption of productive resources can be overwhelming. For example, Hart (1976) estimates that, by the year 2000, costs resulting from hurricanes could average $5 billion (1976 dollars) annually.

This material is based upon research supported by the National Science Foundation under Grant No. CEE 83-09511. Any opinions, findings, and conclusions or recommendations expressed in this publication are those of the author(s) and do not necessarily reflect the views of the National Science Foundation.

Traditionally, when a hurricane threatened a coastal region, the first option was to avert the hazard by relocating inland beyond the threat of the storm and returning after the hazard had abated. However, in the past two decades, the population of coastal regions has increased dramatically. For example, the 1980 population of the Tampa Bay region is projected to almost double by 2000–2010. This increase in coastal and urban population has a direct impact on the vulnerability of residents in such areas.

To evacuate the population at risk in a densely populated area, like the Tampa Bay region, requires well over twenty-four hours (Bastien et al., 1985). However, the U.S. Weather Service cannot predict, within reasonable accuracy, the trajectory of an approaching hurricane beyond 12 hours. Therefore, many evacuation recommendations will result in costly false alarms [e.g., in September 1985, over three-quarters of a million residents evacuated Tampa when threatened by Hurricane Elena, which never struck the area (Tampa Bay Regional Planning Council, 1986)]. If several false alarms occur, some residents may be reluctant to heed such evacuation orders ("the boy who cried wolf" syndrome), but if the hurricane does happen to strike, these "doubters" would have to seek appropriate shelter. Furthermore, if (during the course of a major evacuation) a critical egress route becomes dysfunctional (e.g., the causeway linking Galveston Island to the mainland in Texas), evacuees must seek alternate protection in buildings, or they will be exposed to the direct fury of the storm. Finally, many potential evacuees—particularly the less-mobile, sick, elderly, and handicapped (many located in hospitals and nursing homes)—may be exposed to a greater risk of death or injury by relocating than by remaining in the home facility.

For these and other reasons, people will, in future hurricanes, have to seek shelter in various structures during hurricanes. In some cases, the individual will have to select his shelter. In other cases, local government officials will have designated certain shelters. In either case, the question arises: how safe, during a hurricane, is the occupant in any building?

This chapter and Chapter 4 deal with the safety of occupants in buildings subjected to hurricanes. The chapters are organized into eight major sections:

1. A review of hurricane characteristics
2. A summary of the nonstructural and structural effects of hurricanes on buildings
3. A summary of mitigative design features for structures in hurricane environments
4. A discussion of conventional concepts of occupant safety
5. A description of a more modern approach to occupant safety
6. A model to evaluate occupant safety
7. A demonstration of occupant safety evaluation in a selected structure
8. An evaluation of the effect of building type, hurricane magnitude, and occupant location on the safety of occupants.

Chapter 4 concludes with a section on future research needs in the area of occupant safety in a hurricane environment.

3.2 AN OVERVIEW OF HURRICANE CHARACTERISTICS

Hurricanes are tropical cyclones in which tangential windspeeds equal or exceed 74 mph. Meteorologically, the hurricane is characterized by a strong thermally direct circulation with the rising of warm air near the center of the storm and the sinking of cooler air outside. Air flow at the lower level is directed down the radial pressure gradient, from the higher toward lower pressure, while the outflow takes place at much higher levels, where the radial pressure gradient is very weak. Hurricanes generate and maintain a "warm core" structure without any reliance upon pre-existing heating gradients. The positive temperature anomalies in the interior of the hurricane result from the heating and moistening of the low-level inflow air by the fluxes of water vapor and heat from the highly disturbed sea surface, together with the release of latent heat by deep cumulus convection in the eye-wall cloud and spiral rainbands. The warm core of the hurricane serves as a reservoir of potential energy, which is continuously being converted into kinetic energy by the thermally direct circulation. This energy conversion is responsible for generating the high tangential windspeeds in the storm and maintaining them against frictional dissipation.

The atmospheric pressure gradient in the interior of the storm affects the velocity of the rotational winds. Strong winds, such as those that exist near the center of a hurricane, can exist only in regions with large pressure gradients. The atmospheric pressure in the center of a hurricane must be very low in order to concentrate a rapid decrease in pressure in a short distance. The pressure at the center of a hurricane is, therefore, an indicator of its intensity.

The size and lifespan of a hurricane is also of interest to all concerned. The size of a hurricane may be determined in terms of the diameter in which the windspeed is greater than 74 mph. A typical hurricane may have hurricane-force winds over an area 100 miles in diameter, and gales (above 40 mph) may extend 350–400 miles across. The life of a hurricane may vary from eight to twelve days. The factors determining the lifetime of a hurricane include the time and place of origin, and the general circulation features existing in the atmosphere at the time of occurrence.

All hurricanes that ultimately affect the Gulf and Atlantic coasts of the United States form in the North Atlantic, Caribbean Sea, or Gulf of Mexico areas. While most hurricanes occur in August, September, and October, the six-month period from June 1 to November 30 is defined to be the Atlantic hurricane season. A detailed account of the meteorology of hurricanes can be found elsewhere (Simpson and Riehl, 1981; Wallace and Hobbs, 1977).

3.3 STRUCTURAL AND NONSTRUCTURAL EFFECTS ON BUILDINGS

Depending upon its location, a structure exposed to a hurricane can be subjected to extreme wind loadings and various levels of flooding, scour, surge, and battering by water and airborne debris. At one extreme, in a low elevation coastal setting, the wind velocity is highest, flooding is highly probable, and water could be moving at a significant speed, inducing additional hydrodynamic forces on the structure. In addition, the flowing water transports large floating objects, which can cause significant damage if they should collide with an existing structure. The flowing water also increases the likelihood of scour around foundations, which then renders the building even more susceptible to other environmental forces. At the other extreme, structures located outside high-velocity zones, as defined by the Federal Insurance Administration as Zone V of their

Flood Insurance Maps (Federal Emergency Management Agency, 1984), are subject primarily to winds.

Findings reported by structural engineers performing windstorm damage pathology have identified several failure mechanisms in structures subjected to hurricanes. Severe damage or the total destruction of many structures could be traced to the failure of a building component that constituted a weak link in the structural system. Failures of this type can lead to progressive failure of the major load-resistant components of the structural system. The weak link is very often a support for an overhead door; a connection that does not adequately transmit the forces from one component to another; an unreinforced or inadequately reinforced masonry wall; roof joists or purlins that are inadequately anchored; or horizontal bond beams, in walls, that are not tied into columns or walls with vertical reinforcement. Many other weak links of this type could be mentioned.

The lack of lateral stability (i.e., the ability to resist lateral loads) is also a major cause of failure in the hurricane environment. To provide stability, structural components (in addition to those required to resist gravitational loads) must be introduced. In some structural systems, such as rigid steel frames, no additional material is specifically required to resist wind loads. In regular design practice, the design of wind bracing is a two part operation: first, a suitable arrangement of structural components is determined, and second, forces and moments of these components are used to calculate the members' proportions.

Gross rigid-body motion of the structure in an overturning mode (or a sliding mode) may be induced by wind pressures. These modes of failure are often observed in mobile homes and other unanchored, or poorly anchored, structures.

The stability of the cladding system (walls and openings) is of utmost concern in a hurricane environment. The wind flowing around a building interacts with the geometry of the building to create a complex distribution of pressure on the walls. In a rectangular building, for example, the center of the windward wall sustains relatively high inward pressures, while the pressure reduces near the edges of the windward wall. At the same time, the leeward wall generally experiences outward pressure. However, at the edges of the leeward wall, outward pressure is higher than the pressures recorded at the center of the wall. Side walls normally experience outward pressure, and corners experience relatively large outward pressures as a result of the turbulence in the flow.

The existence of openings in buildings (either existing in a wall or created in the wall due to the failure of windows, doors, or panels) drastically effect the net pressure distribution on the walls and the roof. While the outward or inward sense of the pressure depends upon the location of the opening, the magnitude of the pressures are a function of the size of the opening.

The wind loading, and therefore the stability, of the roof depends upon the pitch of the roof and the relative dimensions of the building. A flat roof experiences an outward pressure due to wind. For a low pitched roof, the direction of wind pressure on the roof depends on the configuration and relative dimensions of the building. A high pitched roof (of, for instance, 50 degrees) may have inward pressure on the windward portion of the roof, regardless of the relative dimensions of the building.

Finally, the loadings caused by eaves, overhangs, sharp edges, and ridges should be identified. Eaves and overhangs of a roof experience entrapped wind underneath them and create turbulence. The combination of these causes lead to significant stagnation pressure on the eaves and overhanging portions of roofs. Sharp edges of roof ridges and wall corners of a building also create turbulence in the wind, which ultimately leads to relatively large negative changes in pressure near the ridges and the corners. A more complete account of the effects of high wind on structural and nonstructural components of buildings may be found elsewhere (Kummer and Sprankle, 1973).

3.4 MITIGATIVE ACTIONS IN HURRICANE-PRONE REGIONS

In the past decade, several authors have addressed the problem of making coastal construction more resistant to hurricane-induced forces. Collier (1978) outlines such preventive measures as entrusting design and construction only to competent professionals; using a 100-year return period for the windspeed design as well as the wave force design; preselecting the location and elevation of the structures to anticipate the hurricane forces; paying special attention to the foundation design such that it performs adequately in the hurricane environment (e.g., to resist undermining); and paying special attention to the superstructure, to ensure that the design loads are resisted via the proper choice of building components and connections.

Saffir (1983) provided a comprehensive treatment of how to develop hurricane-resistant designs *vis-a-vis* the building code of South Florida. In that work, he discussed such topics as the nature of detailed specifications that are required for concrete block masonry type construction, reinforced concrete tie columns, and tie beams. He also discussed preferred standards for foundation design — both slab and pile foundations. The work also included a discussion of steel and timber construction, cladding and glazing, and issues in code enforcement in a hurricane-prone region.

However, from another viewpoint of protection from hurricanes, the buildings and structures subjected primarily to wind loading can be classified into two broad categories: 1) structures for which wind loading can be regarded as quasistatic (these structures are subdivided into two types on the basis of the protective measures required from a) localized high suction, and b) lateral instability); and 2) structures for which the dynamic effects of wind loading have to be taken into account (on the basis of protective measures adopted, these structures fall into three types, such as a) structures sensitive to gust; b) structures subject to vortex excitation; and c) structures under the action of coupled bending and torsional oscillation). Only structures for which the wind can be treated as quasistatic will be discussed here. Details on the dynamic behavior of structures can be found elsewhere (Davenport, 1972; Simiu and Scanlan, 1986).

3.4.1 Protection From Localized High Suction

From the examination of hurricane load damage, and wind tunnel investigations, it has been found that the majority of the damage has been caused by localized high suction, particularly on lightweight roofs and cladding. Furthermore, Dutt (1971) found (from wind tunnel investigations) that the localized high suctions could be as high as $-4q$ (where q is the dynamic pressure of the oncoming wind stream), depending on the direction of the wind flow in relation to the structure, while the average suction on the roof as a whole is on the order of $-q$ only. The localized high suctions are confined to very small areas of the roof and are likely to be effected by minor variations in the eaves details. In the case of roofs with edge beams, such local high suctions are not likely to be very significant as regards the design of the whole roof structure, because the corner is very stiff and could not be lifted without lifting the roof as a whole. However, this condition must be borne in mind while detailing the roof covering.

In the case of roofs without edge beams or similar stiff members, it is absolutely essential to withstand an average local suction of $-2q$ acting up to a width of approximately one-tenth of the span of the roof all along its periphery. The remaining portion of the roof should be fixed down to overcome a suction of $-q$. The additional cost involved in holding down the peripheral zone of the roof to withstand this high average suction of $-2q$ will be small, and may substantially reduce the risk of collapse.

Dutt (1971) has also shown, from wind tunnel investigations, that the average suction on the roof of an open-sided building is very much less than that on the roof of a closed-sided building. From wind tunnel investigation, it would appear adequate to design the walls of the buildings to resist an average wind pressure of q or a suction of $-q$, depending on the wind direction, except for the periphery of the wall. For the peripheral zone of approximately one-tenth the span, an average localized suction of $-2q$ can be assumed. Again, if the building is airtight, it would be adequate to assume a pressure less than q or a suction less than $-q$.

3.4.2 Protection From Lateral Instability

Stability (i.e., the property of withstanding horizontal wind loads) is a requirement that all buildings and structures must meet. This requirement can be achieved, for example, by 1) the provision of diagonal bracing, or 2) by using infill panels, or 3) by making one or more joints rigid. Of the three methods, diagonal bracing is the most efficient and satisfactory structurally, and in the majority of cases it is the most simple and cheapest solution.

Stabilization by the provision of infill panels can be a practical and economical solution, particularly in buildings where walls are required for fire proofing and sound insulation. A line of plane frames may be made stable if even one panel is braced, as long as there is structural continuity throughout the whole length of the frame.

When it is required to keep the walls free of obstructions, the structures are stabilized by making the frame rigid. In the case of reinforced concrete buildings, lateral stability is achieved through the concrete rigid frame, in which integral floor slabs are supported on concrete beams that are framed to the columns. Lateral stability can also be achieved by the provision of shear walls.

3.5 EXISTING CONCEPTS OF OCCUPANT SAFETY

Over the past two decades, several attempts to evaluate the structural performance of existing buildings have been published (Yao, 1979). The highly cited work by Culver et al. (1975) presented methods to evaluate structures subjected to earthquakes, hurricanes, and tornadoes. In that work, three methods of analysis, each distinguished by the complexity of the structure or its intended use, were proposed. In the first method, the "Field Evaluation Method," buildings were evaluated qualitatively on the basis of their structural characteristics, structural configuration, and the observed degree of deterioration. The intent was to provide a rapid, inexpensive means of identifying hazardous or potentially hazardous structures.

In the second method, the "Approximate Analytical Evaluation Method," buildings were evaluated on the basis of the behavior of anticipated critical structural members. Using information from design and construction documents, as well as anticipated loads on the structure provided by codes, an elastic-static structural analysis was performed to identify the critical structural members.

The third method, the "Detailed Analytical Evaluation Method," computed the damage level in the structure subjected to a wind hazard. Damage was related to the story ductility (i.e., calculated interstory drift of the ith story divided by a user-specified interstory drift-to-yield factor). A versatile computer program was provided with the report. To use the Detailed Analytical Method requires specific information about the structural properties and geometry of the building and the loading on the structure. The authors intended the procedure to be used for complex or critical structures, such as hospitals and communication centers.

One stated purpose of the three methods for evaluating existing buildings was to determine the risk to life safety under natural hazard conditions. The authors claimed that, while the safety of the building occupants cannot be evaluated directly, the safety of the occupants can be related to the structural performance and the resulting damage to the building. Without further discussion on how building damage may be quantitatively related to occupant safety, the remainder of the report dwelled on the evaluation of damage. Although the authors did not specify an acceptable level of damage, they pointed out that such a specification varied with the usage and function of the structure. They recommended that the interpretation of the level of damage predicted by the proposed methods 'as they relate to life and property loss' be exercised by the user.

Hasselman et al. (1980) developed a computer program to assist building and safety officials in calculating the damage potential to multistory buildings exposed to earthquake, severe wind, and tornado forces. The assumption in this work, as in the Culver study, is that the potential safety of the occupants is related to the damage. Building interstory drift was again taken as the indicator of damage. The determination of damageability characteristics of the building components was based on expert opinion and a limited amount of data. The final damage to the structure was reported as a percentage of replacement cost on a floor-by-floor basis. Obviously, a city official, in making any safety decision, must take on the responsibility of relating building damage level to occupant safety.

More recently, Mehta et al. (1981) proposed a procedure for predicting wind damage to buildings. Two procedures, one subjective and the other analytical, were used to evaluate potential windstorm damage to existing buildings. In the subjective approach, an on-site survey of the building is performed to establish structural details. The resulting damage to the structure is inferred by using damage experience from similar structures. In the analytical approach, a structural analysis is performed, based on a knowledge of the prevailing aerodynamic forces and the strength of the building components used in the structure. Results of the analysis provided a scenario of the sequence of damage to the structure as a function of windspeed. The authors also claimed that, by using a wind-hazard probability model and the results of the deterministic structural analysis, the probability of damage sequence can be determined.

Recently, Spangler and Jones (1984) were among the first to address the specific problem of structural certification of potential hurricane shelters for vertical evacuation (i.e., the use of deliberately-selected multistory structures as shelters). The authors proposed the following procedure to evaluate a structure: 1) identify potential shelters; 2) collect relevant information on the structure; 3) physically inspect the structure; 4) physically inspect the surrounding terrain; 5) analyze the collected information; and 6) rate the safety of the building. The safety of the structure was based on either a subjective opinion or a static structural analysis of the building. The resistance of the structure was expressed in terms of the magnitude of the hurricane, expressed in terms of the Saffir-Simpson scale, that the structure could 'safely' withstand. The decision that a structure could withstand a hurricane of a given magnitude was based on the allowable stress and stability criteria defined by existing building codes.

3.6 A MORE RATIONAL APPROACH TO OCCUPANT SAFETY

Recent developments in structural mechanics have shown that there is no rational explanation of the degree of absolute safety provided by structures designed using traditional working stress design and ultimate strength methods (ASCE, 1972; Freudenthal et al., 1966; and Thoft-Christensen and Baker, 1982). In response to these and subsequent findings, the current trend has been to take into consideration the random nature of the loads to which structures are subjected and the variations in the material properties of the structural constituents. In other words, the loads impacting a structure, and the resistance of that structure, are considered to be random variables. The safety margin provided by the structure is the amount by which the random resistance of the structure exceeds the random load applied to the structure. Failure is said to occur when the safety margin is less than, or equal to, zero. The relative safety of a structure is now expressed in terms of a probability of failure. The smaller the probability of failure, the safer the structure.

One advantage of this more modern approach is the realization that safety is a relative concept. But the perennial question — 'how safe is safe?' — remains unanswered (Burton and Pushchak, 1984; Derby and Keeney, 1981). One school of thought defines safety in terms of levels of risks (Lowrance, 1976). Thus, what is safe depends upon what levels of risk an individual is willing to take on a voluntary basis, or the levels of acceptable risk that a governmental agency may set (Starr, 1969). The latter risk level, established by government agency, is sometimes referred to as the involuntary risk. In both cases, the individual or the governmental decision will depend upon the extent to which alternatives are available. The voluntary or the involuntary safety levels for occupant safety may depend, therefore, upon specific scenarios and competing alternatives.

In the last section we found that, in most of the traditional approaches aimed at evaluating the safety of occupants in existing buildings, the protection offered by the structure was evaluated on the basis of either the level of damage sustained by the structure, or the extent to which a structure satisfied a particular building code. For the same loading environment, one of two structures is considered safer if 1) the factors of safety of its elements are larger; or 2) the predicted damage sustained by the structure is smaller. In all cases, the evaluatory criteria for structural safety are tied to the structure. However, such criteria may not guarantee the safety of the occupants of the structure. For example, an occupant may be injured or killed as the result of a falling ceiling or a collapsing partition. Furthermore, factors of safety for structural elements or the probabilities of failure of a structural frame may not have the same interpretation for different structures. It is conceivable, for example, that two different structures (say a ductile steel structure and a brittle masonry structure) having identical failure probabilities, or experiencing the same magnitude of damage, may result in different levels of injury or death to occupants.

Since, in any structure subjected to a hurricane, the potential for injury resulting from nonstructural causes may be equal to, or greater than, that resulting from structural failure, it is fitting to propose a method of structural evaluation that focuses directly on the safety of the occupant, and which simultaneously accounts for the environmental forces, the structural characteristics, and the nonstructural failure characteristics of the structure. It is felt that the risk of injury or death to an occupant of a structure is a rational measure of occupant safety: the smaller this risk, the greater level of occupant safety offered by the structure. What is needed is a logically consistent and practical method of evaluating these risks.

3.7 A MODEL TO EVALUATE OCCUPANT SAFETY

3.7.1 A Fault-Tree Model

It has been shown elsewhere (Stubbs and Sikorsky, 1985) that fault-tree analysis can provide a rational and conceptual framework to evaluate the safety of occupants in a structure exposed to a hurricane. The fault-tree analysis process starts with a defined 'undesired' event (i.e., the top event), then proceeds by deduction to develop a set of contributory events that can cause the top event. The process is continued for each of the contributory events until the resulting contributory events become basic events (i.e., events for which statistical information is readily available or can be developed by analysis). The method generates a diagram (called a fault tree) that is a model of the event relationships for the system (Barlow and Lambert, 1975). A description and definition of the symbols used in developing the model are provided in Figures 3.1 and 3.2. This

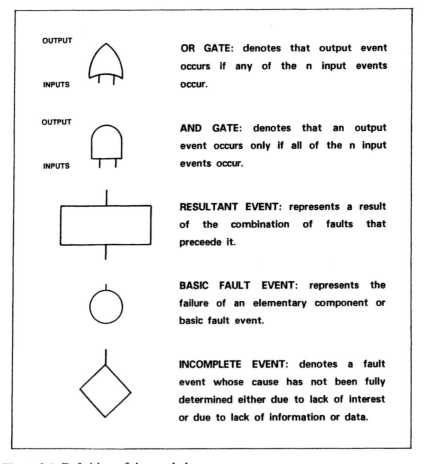

Figure 3.1. Definition of the symbols.

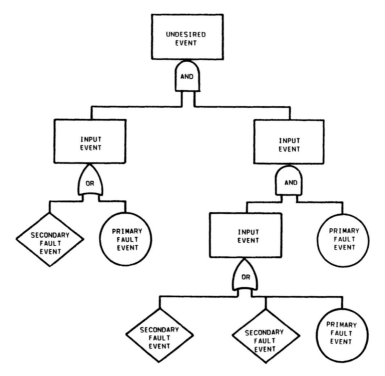

Figure 3.2. Typical fault tree.

method has been used in such diverse applications as nuclear power plants (Cummings, 1975; Rasmussen, 1974), the safety analysis of piping systems (Abes et al., 1985), and the reliability analysis of construction field instrumentation (Kuroda and Miki, 1985).

The system of interest in this study is any structure subjected to a hurricane, and in which occupants are taking shelter. The structural and nonstructural elements of the building include the foundation, the structural framing system to transfer the loads to the foundation, exterior walls or cladding, openings in the exterior walls (doors and windows), the roof, internal partitions, and floors, among others (e.g., the mechanical system). In this study, the undesired event is a fatality or an injury that occurs during the course of the hurricane. However, since exactly what constitutes an injury may be difficult to define, the top event will be limited to potential human fatalities.

The fault tree presented in Figure 3.3 represents a comprehensive

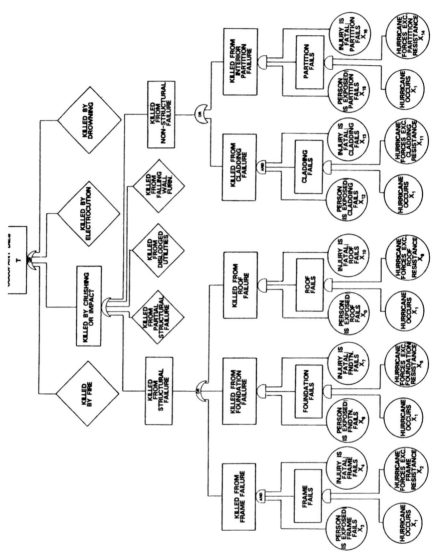

Figure 3.3. Fault-tree model for occupant safety.

model that relates the basic fault events to occupant safety. The model contains several attributes. First, this model of occupant safety is general (i.e., the same formulation can be applied to many structural types with little or no modification). Second, the model is highly integrative (i.e., it pulls together the occurrence of structural, as well as nonstructural, failures). As will be shown later, it also allows a smooth interface between existing methods of safety evaluation, such as reliability analysis and occupant safety. The model also integrates the occurrence of other hazards (e.g., water forces, windborne debris) that may simultaneously occur during a hurricane. Third, the model is comprehensive. Assuming that data are available, the relative importance of each hazard type may be determined. Put another way, the model clearly states what information is needed to perform an analysis of occupant safety.

The objective of a quantitative analysis is to determine the probability of the occurrence of the top event. Using the basic events defined in Figure 3.3 and listed in Table 3.1, the following new events are defined:

$$
\begin{aligned}
Y_1 &= X_2 \cap X_3 \cap X_4 & Y_2 &= X_5 \cap X_6 \cap X_7 \\
Y_3 &= X_8 \cap X_9 \cap X_{10} & Y_4 &= X_{11} \cap X_{12} \cap X_{13} \\
Y_5 &= X_{14} \cap X_{15} \cap X_{16}
\end{aligned}
\qquad 3.1
$$

Then the top event T is given by:

$$
\begin{aligned}
T = (X_1 \cap Y_1) \cup (X_1 \cap Y_2) \cup (X_1 \cap Y_3) \\
\cup (X_1 \cap Y_4) \cup (X_1 \cap Y_5)
\end{aligned}
\qquad 3.2
$$

Using the distributive law, the repeated event X_1 can be eliminated, to give:

$$
T = X_1 \cap (Y_1 \cup Y_2 \cup Y_3 \cup Y_4 \cap Y_5)
\qquad 3.3
$$

from which the probability of the top event, P, is now given by:

$$
\begin{aligned}
P(T) = P(X_1)[P(Y_1) + P(Y_2) + P(Y_3) + \\
P(Y_4) + P(Y_5) + \ldots]
\end{aligned}
\qquad 3.4
$$

in which,

Table 3.1 Definition of basic events.

EVENT	DESCRIPTION OF EVENT
T	Occupant dies
X_1	Hurricane occurs
X_2	Hurricane forces exceed frame resistance
X_3	Person is exposed|frame fails
X_4	Injury is fatal|frame fails and person is exposed
X_5	Hurricane forces exceed foundation resistance
X_6	Person is exposed|foundation fails
X_7	Injury is fatal|foundation fails and person is exposed
X_8	Hurricane forces exceed roof resistance
X_9	Person is exposed|roof fails
X_{10}	Injury is fatal|roof fails and person is exposed
X_{11}	Hurricane forces exceed cladding resistance
X_{12}	Person s exposed|cladding fails
X_{13}	Injury is fatal|cladding fails and person is exposed
X_{14}	Hurricane forces exceed interior partition resistance
X_{15}	Person is exposed|partition fails
X_{16}	Injury is fatal|partition fails and person is exposed

$$P(Y_1) = P(X_2)P(X_3)P(X_4) \qquad P(Y_2) = P(X_5)P(X_6)P(X_7) \qquad 3.5$$
$$P(Y_3) = P(X_8)P(X_9)P(X_{10}) \qquad P(Y_4) = P(X_{11})P(X_{12})P(X_{13})$$
$$P(Y_5) = P(X_{14})P(X_{15})P(X_{16})$$

The engineering effort is now focused on determining the probabilities of occurrences of the basic events, and using Equation 3.4 to estimate the probability of occurrence of the top event.

3.7.2 Risk Models for Occupant Safety

If N is the number of people sheltered in the structure, the risk associated with using the structure, $E[N]$ (the expected number of fatalities), may then be estimated using the equation:

$$E[N] = P[T]N \qquad\qquad 3.6$$

where $P[T]$ is given in Equation 3.4. Some investigators, however, find the definition of risk in Equation 3.6 to be lacking in some respects. For example, if, for two alternative options, a and b, $P[T]_a N_a = P[T]_b N_b$, where, $P[T]_a \gg P[T]_b$, how does one decide between them on the basis of expected risk? One way to overcome this limitation is to represent risk via so-called risk curves (Kaplan and Garrick, 1981) or risk profiles (Fiksel and Rosenfield, 1982). The risk curve, denoted by $R(x)$, is the complement of the probability distribution function of the annual losses, and is defined to be:

$$R(x) = P[\text{annual losses exceed } x] \qquad\qquad 3.7$$

where x is some realization of the random variable X.

Closely associated with the risk curve above is the "conditional risk profile," which, for a specific hazard E, is defined to be:

$$R(X|E) = P[\text{losses exceed } x \text{ given that event } E \text{ occurred}] \qquad 3.8$$

Note that the expectation of the losses $E[X]$: may be given by:

$$E[X] = \int_{-\infty}^{\infty} x[d\{1 - R(x)\}/dx]dx \qquad\qquad 3.9$$

which can be shown to be the area under the curve (Rowe, 1977). Similarly, the variance of the losses may be computed from

$$\text{Var}[X] = \int_{-\infty}^{\infty} (x - E[X])^2[d\{1 - R(x)\}/dx]dx \qquad\qquad 3.10$$

If we define the random variable X to be the fraction of the building inhabitants to be killed, and the event E to be a hurricane of some predetermined intensity, then the fault-tree model may be used to generate risk curves by making the substitutions shown in Table 3.2. For each hurricane, the realization of the random variable X, x_i, will take on values between zero and one.

Thus, the protection offered by a structure to the occupants is described by the two moments, $E[X]$ and $\text{Var}[X]$, which depend, in turn, upon the structural and nonstructural characteristics of the

Table 3.2 Definition of events.

OLD EVENT	NEW DESCRIPTION OF EVENT
T	$(X > x_i)$ |Hurricane occurs
X_4	$(X > x_i)$ |Frame fails and hurricane occurs
X_7	$(X > x_i)$ |Foundation fails and hurricane occurs
X_{10}	$(X > x_i)$ |Roof fails and hurricane occurs
X_{13}	$(X > x_i)$ |Cladding fails and hurricane occurs
X_{16}	$(X > x_i)$ |Partitions fail and hurricane occurs

building, the magnitude of the hurricane, the exposure of the occupants, and the consequences of failure of any of the structural or nonstructural components.

3.7.3 Probability Assignment for Basic Events

The events needed for input into the present model may be classified into four categories: statistics describing the hurricane (X_1), statistics describing the reliability of the building components $(X_2, X_5, X_8, X_{11}, X_{14})$, statistics describing the exposure $(X_3, X_6, X_9, X_{12}, X_{15})$, and statistics describing the consequences resulting from the failure of the building components $(X_4, X_7, X_{10}, X_{13}, X_{16})$.

In evaluating a given structure, it will be assumed that the structure receives the full force of the given hurricane. Thus, $P[X_1] = 1$.

The conditional failure probabilities of the building components $P(X_2)$, $P(X_5)$, $P(X_8)$, $P(X_{11})$, and $P(X_{14})$ (i.e., the frame, foundation, roof, cladding, doors and windows, and partitions, respectively) will be developed using established techniques from the field of structural reliability.

Since the structures considered are assumed to be occupied during a hurricane, people will be exposed if failure of any of the subsystems occurs. Thus, we may set:

$$P[X_3] = P[X_6] = P[X_9] = P[X_{12}] = P[X_{15}] = 1 \qquad 3.11$$

Note that these numbers may vary, depending on the exact location of the occupant.

Whereas analytical techniques exist that permit the calculation of failure probabilities for structures, we know of no comparable set of techniques that would permit us to estimate the consequences of failure in terms of fatalities or injuries. Therefore, until such techniques are available, we must resort to empirical means. If we assume that the relationship between damage and fatalities is independent of the force system that caused the damage, then we may use the earthquake-generated damage-fatality data to predict fatalities in a hurricane. A summary of fatality statistics for various classes of structures damaged at various levels is presented in Table 3.3 (Anagnostopoulos and Whitman, 1977; Whitman et al., 1980). Until better data are made available, we will be conservative and assume that if any of the designated protective subsystems fail (i.e., roof, cladding, frame, or foundation), the consequences of failure corresponding to frame collapse will result.

3.8 OCCUPANT SAFETY EVALUATION FOR A TYPICAL SHELTER

The analysis procedure consists of the following steps: 1) a description of the structure and the surrounding terrain; 2) a description of the sources of information; 3) a description of the analysis procedure; 4) a definition of failure of the building subsystems; 5) a determination of resistance and loading statistics; and 6) the risk analysis.

Table 3.3 Mean values and standard deviation for fatalities as a fraction of total occupants.

| DAMAGE STATE | WOOD | MASONRY | | REINF. CONCRETE; STEEL | |
		RESIDENTIAL	COMMERCIAL	≤ 5 STORIES	> 5 STORIES
Moderate	0.0001[a]	0.0015	0.0005	0.00012	0.00012
	(0.0013)[b]	(0.0013)	(0.0028)	(0.0011)	(0.0011)
Heavy	0.001	0.002	0.0045	0.0016	0.0016
	(0.0057)	(0.0079)	(0.016)	(0.0071)	(0.0071)
Total	0.008	0.018	0.0245	0.021	0.024
	(0.024)	(0.038)	(0.048)	(0.041)	(0.043)
Collapse	0.07	0.2	0.4	0.4	0.6
	(0.23)	(0.25)	(0.41)	(0.41)	(0.44)

Mean fraction of fatalities.
Standard deviation.

3.8.1 Description of the Structure

The structure, located on the Gulf Coast, serves as a retirement home and a health-care facility. The waterfront structure, built in 1964–1965, consists of a seven-story main structure and a two-story health-care facility. Each floor in the main structure covers an area of 13,000 square feet, while the health-care facility covers an area of approximately 6,500 square feet per floor. The height of the first story is 12.0 feet above ground level. The height of each succeeding floor is 9.67 feet.

Figure 3.4 shows a typical floor plan and elevation of the structure. The foundation consists of footings supported on piles (three or four per column, with a resistance of 45 tons per pile). The lateral load-resisting system consists of a moment-resisting reinforced concrete frame. The cladding is comprised of masonry walls, and windows using $7/32$-inch heavy sheet glass. In the main structure, west, southeast (short), southeast (long), and northeast elevations have 384, 18, 245, and 0 windows, respectively. The two-story part of the building has 60 and 28 windows on the southeast (long) and northeast sides, respectively. The floors and the roof are of monolithic-type reinforced concrete construction. Roofing consists of built-up tar and gravel.

3.8.2 Sources of Building Information

Data defining the structure were obtained from three sources: 1) construction plans and specifications for the building; 2) the Standard Building Code, the AISC Specifications for Design, Fabrication, and Erection of Structural Steel for Buildings, and ACI 318; and 3) a visual inspection of the structure and the surrounding terrain.

3.8.3 Analysis Procedure

The key steps of the evaluation procedure were as follows: 1) failure functions for structural units (cladding elements, doors, windows, roof elements, etc.) were defined (to contain the complexity of the analysis, linear failure functions were selected where possible); 2) loading statistics (derived from the hurricane) and resistance statis-

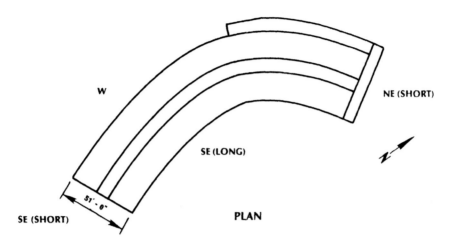

W

NE (SHORT)

SE (LONG)

N

SE (SHORT)

51' - 0"

PLAN

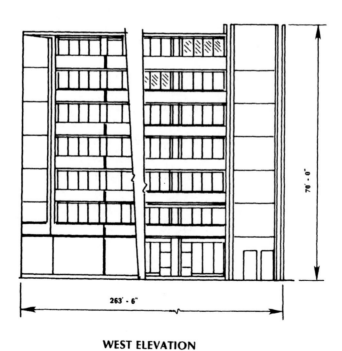

70' - 0"

263' - 6"

WEST ELEVATION

Figure 3.4. Plan and elevation of existing example structure.

tics (derived from the materials and the design specifications) were determined for the unit; 3) approximate failure probabilities for the units were defined, using Mean Value Methods from Structural Reliability Theory; 4) failure scenarios for the building subsystems were synthesized, using Fault-tree Analysis; 5) failure probabilities for the building subsystems were computed; 6) risks of fatalities were computed, using Equation 3.4; 7) modifications were made to upgrade the structure to resist a Category 3 hurricane; and 8) new risks of fatalities and costs to upgrade the structure were computed.

3.8.4 Working Definitions of Building Subsystem Failure

On the basis of historical observation, the chances that the frame or foundation of a professionally engineered structure will fail in a Category 1, 2, or 3 hurricane is quite small. In fact, no such failures have been documented for structures subjected primarily to wind. If a structure is subjected to wind and water, the cladding will usually fail first, thus relieving the load on the frame. If the foundation is well constructed (e.g., on piles), failure by scour can also be ruled out. Therefore, in this study, if the structure was professionally engineered, and the foundation was designed to resist scouring, frame and foundation failure were ignored.

Furthermore, if the above restrictions are in effect, in Categories 3–5 hurricanes, the chances of roof and cladding failures are many times greater than the chances of frame or foundation failure. Thus, if the consequences of roof and cladding failure are of the same order of magnitude as the consequences of frame and foundation failure, the risks of frame and foundation failure are small compared to the risk associated with roof and cladding failure. Therefore, frame failure may be neglected in the risk analysis for Categories 3–5.

The roofing system is modelled as a two-element series system, consisting of the roof beams and the roof deck. Failure of the roof system occurs if a major structural unit supporting the deck (e.g., beam/joist) fails, or if more than five percent of the deck area fails. To determine deck failure, the roof decking was divided into equal panel sizes and the failure probability of one panel determined. The failure characteristics of the deck system was estimated by assuming that the failure of each panel was independent, and that the failure characteristics of the system could be modelled by a binomial distribution. The probability of failure of one panel was equated with the probability of a "success" in the binomial sense.

Cladding failure occurs if the cladding system provides no protection from external hazards. Operationally, failure of the cladding system occurs if more than 40 percent of the cladding area of each wall on opposite sides of the building is lost. The building cladding area was divided into equal panel sizes and the reliability of one panel determined. Using binomial distribution, as above, the failure probability of each side was determined. Assumed panel sizes for the roof and cladding systems were based on engineering judgment. Factors considered included: the type of construction material, spacing of cladding supports, and spacing of decking supports.

3.8.5 Determination of Loading and Resistance Statistics

Hurricane categories were identified by the Saffir-Simpson Scale (Simpson and Riehl, 1981). The mean and variance of the windspeed for each hurricane category were estimated, assuming a uniform probability density function. For example, if V_a and V_b are the lower and upper values of windspeed for a given category, the mean speed is $\dfrac{(V_a + V_b)}{2}$ and the variance is $\dfrac{(V_b - V_a)}{12}$.

Wind pressure acting on the structural frame was determined by increasing the basic wind pressure [$P = 0.00256V^2(H/30)^{2/7}$] by a shape factor of 1.3 ($C_D = 1.3$). To model the hurricane wind field, the wind pressure above a height of 30 feet acting on the building was assumed constant. From ground level to a height of 30 feet, wind pressure was assumed to vary linearly. Statistics for the windspeed and lateral pressures resulting from the five hurricane categories (See Appendix A) are listed in Table 3.4.

Wind pressures acting on the roof, cladding, and openings were determined by increasing the basic wind pressure by a shape factor of 1.5 ($C_D = 1.5$). The statistics describing the representative uplift pressures on the roof are listed in Table 3.5.

Water force calculations were based on procedures provided in *Shore Protection Planning and Design* (Coastal Engineering Research Center, 1966). Results of a SLOSH analysis for the site indicated that, for the given site, surge depths were zero for Hurri-

Table 3.4 Hurricane categories and their resulting loading on the structures.

HURRICANE[A] CATEGORY (*i*)	WINDSPEED V_i(MPH)		WIND PRESSURE P_i(PSF)	
	MEAN	VARIANCE	MEAN	VARIANCE
1	84.5	12.25	23.76	9.52
2	103.0	5.44	35.31	15.02
3	120.5	10.03	48.32	29.80
4	143.0	16.00	68.05	60.81
5	165.5	10.03	91.15	95.26

[A]See Appendix A.

cane Categories 1 and 2; and 7.7 feet, 12.3 feet, and 14.2 feet for Hurricane Categories 3, 4, and 5, respectively.

The nominal design resistance (R_{design}) for the various structural elements was assumed to be the allowable loads listed on the construction drawings. If information regarding the design resistance was not available, it was estimated from the applicable building code. Mean resistances for cladding and roof elements were estimated as follows (see Appendix B):

$$\overline{R} = R_{design}/(1 - \beta(COV[R]))   3.12$$

where β is the reliability index associated with the design, and COV[R] is the coefficient of variation associated with the parame-

Table 3.5 Hurricane uplift pressures[a] (psf).

HURRICANE CATEGORY	MEAN	COV
1	29.25	0.17
2	43.45	0.13
3	59.47	0.16
4	83.76	0.13
5	112.19	0.12

[a]$P = 0.00256C_D V^2$, $\mu_{C_D} = -1.5$, COV[C_D] = 0.1.

Table 3.6 Reliability indices assumed in study.

BUILDING COMPONENT	P_f	β
Roof elements	0.001	3.093
Wall unit	0.01	2.331
Window or door unit	0.01	2.331

ters under study. Note that β may be estimated from the observed failure rate, P_f, of the various elements using the equation:

$$-\beta = \Phi^{-1}(P_f) \qquad 3.13$$

where Φ is the standard normal distribution.

Values of β used in this study are listed in Table 3.6. Resistance statistics (mean and variance) were also estimated, based on the values given in Table 3.7.

3.8.6 Risk Analysis

Using these definitions of failure for the various building components, and the guidelines for defining the loading and resistance statistics, failure probabilities were computed for frame failure, foundation failure, cladding failure, and roof failure. Detailed failure functions and statistical parameters for the loading and resistance variables can be found elsewhere (Stubbs, 1987). In addition, since stairwell space in the structure under consideration was limited, the exposure probabilities were set at a maximum (i.e., unity),

Table 3.7 Additional resistance statistics used in study.

COMPONENT	$\bar{R}^a/R_N$	COV	ASSUMED DISTRIBUTION	SOURCE
Cold formed steel members	1.17	0.17	Normal	A58.1[b]
Masonry	1.05	0.10	Normal	Engineering judgment
Reinforced concrete	1.22	0.16	Normal	A58.1[b]

[a]$\bar{R}$ = Mean resistance.
R_n = Nominal resistance.
[b]See Ellingwood et al., 1980 and American National Standard, 1982.

as defined in Equation 3.11. These input values are summarized in Table 3.8, for the existing structure, and Table 3.9, for the structure upgraded to resist a Category 3 hurricane. Note that the hypothetical upgrading in this case included only modifications to the cladding and roof. Note also, in Table 3.9, that the weak link in the

Table 3.8 Risk model data input for existing example structure.

	BASIC EVENT PROBABILITIES				
BASIC	HURRICANE CATEGORY				
EVENT	1	2	3	4	5
X_1	1.00	1.00	1.00	1.00	1.00
X_2	a	a	a	a	a
X_3	1.00	1.00	1.00	1.00	1.00
X_5	a	a	a	a	a
X_6	1.00	1.00	1.00	1.00	1.00
X_8	a	1.41×10^{-5}	4.90×10^{-2}	9.99×10^{-1}	1.00
X_9	1.00	1.00	1.00	1.00	1.00
X_{11}	a	a	a	1.00	1.00
X_{12}	1.00	1.00	1.00	1.00	1.00
X_{14}	a	a	a	a	a
X_{15}	1.00	1.00	1.00	1.00	1.00

*Less than 10^{-7}, or failure ignored.

Table 3.9 Risk model data input for upgraded example structure.

	BASIC EVENT PROBABILITIES				
BASIC	HURRICANE CATEGORY				
EVENT	1	2	3	4	5
X_1	1.00	1.00	1.00	1.00	1.00
X_2	a	a	a	a	a
X_3	1.00	1.00	1.00	1.00	1.00
X_5	a	a	a	a	a
X_6	1.00	1.00	1.00	1.00	1.00
X_8	a	a	a	2.58×10^{-3}	1.00
X_9	1.00	1.00	1.00	1.00	1.00
X_{11}	a	a	a	a	1.00
X_{12}	1.00	1.00	1.00	1.00	1.00
X_{14}	a	a	a	a	a
X_{15}	1.00	1.00	1.00	1.00	1.00

*Less than 10^{-7}, or failure ignored.

building system is the roof (Event X_8), which exhibits a relatively significant failure probability (4.9×10^{-2}) for a Category 3 hurricane.

The probabilities associated with the consequences of failure are listed in Table 3.10. These numbers have been extrapolated from the damage-death statistics for a similar structure (reinforced concrete structure, greater than seven stories) subjected to earthquakes. Because the failure of either the roof, cladding, frame, or foundation exposes the occupant directly to the hurricane environment, the consequences of these types of failure were set equal to the consequences of failure of the structure in an earthquake. How these assumptions may be refined is a subject of future research. The results of the analyses are shown in Tables 3.11 and 3.12. Tables 3.11 and 3.12 represent the output of the risk model, Equation 3.4.

Table 3.10 Risk model fatality input for example structure.

BASIC EVENT	DESCRIPTION OF BASIC EVENT	MEAN	STANDARD DEVIATION	DISTRIBUTION
X_4	Fraction x killed|frame fails	0.60	0.44	Lognormal
X_7	Fraction x killed|foundation fails	0.60	0.44	Lognormal
X_{10}	Fraction x killed|roof fails	0.60	0.44	Lognormal
X_{13}	Fraction x killed|cladding fails	0.60	0.41	Lognormal
X_{16}	Fraction x killed|partition fails	0.02	0.04	Lognormal

Table 3.11 Risk of using example structure in various hurricanes.

HURRICANE CATEGORY	EXPECTED FRACTION OF FATALITIES	STANDARD DEVIATION OF EXPECTED FATALITIES
1	a	a
2	3.86×10^{-6}	1.48×10^{-3}
3	1.34×10^{-2}	8.64×10^{-2}
4	4.22×10^{-1}	3.05×10^{-1}
5	4.22×10^{-1}	3.05×10^{-1}

aProbability of failure less than 10^{-7}.

**Table 3.12 Risk of using upgraded example structure
in various hurricanes.**

HURRICANE CATEGORY	EXPECTED FRACTION OF FATALITIES	STANDARD DEVIATION OF EXPECTED FATALITIES
1	a	a
2	a	a
3	a	a
4	7.05×10^{-4}	2.00×10^{-2}
5	4.22×10^{-1}	3.05×10^{-1}

ªProbability of failure less than 10^{-7}.

REFERENCES

Abes, A. J., Salinas, J. J., and Rogers, J. T. Risk assessment methodology for pipeline systems. *Structural Safety*, Vol. 2, No. 3, January 1985, pp. 225–237.

American National Standard, SI, 1982. *Building Code Requirements for Minimum Design Loads in Buildings and Other Structures, ANSI A58.1.* New York: American National Standards Institute, 1982.

Anagnostopoulos, S. A. and Whitman, R. V. On human loss prediction in buildings during earthquakes. *Proceedings of the Sixth World Conference on Earthquake Engineering*, Vol. 1. New Delhi, 1977, pp. 671–676.

Task Committee on Structural Safety of the Administrative Committee on Analysis and Design of the Structural Division. Structural safety—A literature review. *Journal of the Structural Division, ASCE.* Vol. 98, No. ST4, April 1972, pp. 845–884.

Barlow, R. E. and Lambert, H. E. Introduction to fault tree analysis. In *Reliability and Fault Tree Analysis Theoretical and Applied Aspects of System Reliability and Safety Assessment*, Barlow, R. E., Fussell, J. B., and Singpurwalla, N. D., eds. Philadelphia: Society for Industrial and Applied Mathematics, 1975, pp. 7–35.

Bastien, M. C., Dumas, M., Laporte, J., and Parmentier, N. Evacuation risks: A tentative approach for quantification. *Risk Analysis*, Vol. 5, No. 1, March 1985, pp. 53–61.

Burton, I. and Pushchak, R. The status and prospects of risk assessment. *Geoforum*, Vol. 15, No. 3, 1984, pp. 463–475.

Coastal Engineering Research Center. *Shore Protection Planning and Design*, 3rd ed. Washington, D.C.: U.S. Army Corps of Engineers, 1966.

Collier, C. A. Guidelines for beachfront construction in Florida. *Ocean Engineering*, Vol. 8, 1978, pp. 309–320.

Culver, C. G., Lew, H. S., Hart, G. C., and Pinkham, C. W. Natural hazards evaluation of existing buildings. *NBS Building Science Series Report No. 61.* Washington, D. C.: Center for Building Technology, Institute for Applied Technology, National Bureau of Standards, January, 1975.

Cummings, G. E. Application of the fault tree technique to a nuclear reactor containment system. In *Reliability and Fault Tree Analysis Theoretical and Applied Aspects of System Reliability and Safety Assessment*, Barlow, R. E., Fussell, J. B., and Singpurtwalla, N. D., eds. Philadelphia: Society for Industrial and Applied Mathematics, 1975, pp. 805–825.

Davenport, A. G. Wind loading and wind effects. Theme Report, Technical Committee No. 7. Bethlehem, PA: International Conference on Planning and Design of Tall Buildings, Leigh University, 1972.

Derby, S. L. and Keeney, R. L. Risk analysis: Understanding 'How safe is safe enough?'". *Risk Analysis*, Vol. 1, No. 3, September 1981, pp. 217–224.

Dutt, A. J. Wind pressure distribution on hyperbolic paraboloid shell roofs. *Civil Engineering and Public Works Review*. London, Vol. 66, 1971, pp. 65–70.

Ellingwood, B., Galambos, T. V., MacGregor, J. G., and Cornell, C. A. Development of a probability based load criterion for American National Standard A58: Building Code Requirements for Minimum Design Loads in Buildings and Other Structures. *NBS Special Publication 577*. Washington, D.C.: U.S. Department of Commerce, June 1980.

Federal Emergency Management Agency. Flood insurance rate maps, City of Galveston, Texas, Galveston County. *National Flood Insurance Program*, 1984.

Fiksel, J. and Rosenfield, D. B. Probabilistic models for risk assessment. *Risk Analysis*, Vol. 2, No. 1, March 1982, pp. 1–8.

Freudenthal, A. M., Garrelts, J. M., and Shinozuka, M. The Analysis of structural safety. *Journal of the Structural Division, ASCE*. Vol. 92, No. ST1, February 1966, pp. 267–325.

Hart, G. C. *Natural Hazards: Tornado, Hurricane, Severe Wind Loss Models*. Redondo Beach, CA: J. H. Wiggins Company, 1976.

Hasselman, T. K., Eguchi, R. T., and Wiggins, J. H. *Assessment of Damageability for Existing Buildings in a Natural Hazards Environment*. Technical Report No. 80-1332-1. Redondo Beach, CA: J. H. Wiggins Co., September 1980.

Kaplan, S. and Garrick, B. J. On the quantitative definition of risk. *Risk Analysis*, Vol. 1, No. 1, March 1981, pp. 11–27.

Kummer, R. E. and Sprankle, R. B., eds. *Multi Protection Design*, TR-20-(Vol. 6). Washington, D.C.: Defense Civil Preparedness Agency, December 1973.

Kuroda, K. and Miki, S. Reliability assessment of field instruments based on F.T.A. *Proceedings of the fourth International Conference on Structural Safety and Reliability*, Vol. III, I. Konishi, I., Ang, A. H. -S., and Shinozuka, M., eds. Japan: Shinto Publishing, 1985, pp. 373–382.

Lowrance, W. W. *Of Acceptable Risk Science and the Determination of Safety*. Los Altos, CA: William Kaufman Inc., 1976.

Mehta, K. C., McDonald, J. R., and Smith, D. A. Procedure for predicting wind damage to buildings. *Journal of the Structural Division, ASCE*. Vol. 107, No. ST11, November 1981, pp. 2089–2096.

Rasmussen, N. C. Reactor safety study: An assessment of accident risks in U.S. commercial nuclear power plants. *Wash-1400*. Washington, D.C.: USAEC, August 1974.

Rowe, W. D. *An Anatomy of Risk*. New York: John Wiley & Sons, 1977.

Saffir, H. S. Practical aspects of design for hurricane-resistant structures; Wind loadings. *Journal of Wind Engineering and Industrial Aerodynamics.* Vol. 11, 1983, pp. 247–259.

Simiu, E. and Scanlan, R. H. *Wind Effects on Structures,* 2nd ed. New York: John Wiley & Sons, 1986.

Sompson, R. H. and Riehl, H. *The Hurricane and Its Impact.* Baton Rouge: Louisiana State University Press, 1981.

Spangler, B. D. and Jones, C. P. Evaluation of existing and potential hurricane shelters. *Sea Grant Project No. R/C - 9.* Gainesville, FL Florida Sea Grant College, November 1984.

Starr, C. Social benefit versus technological risk. *Science.* Vol. 165, September 19, 1969, pp. 1232–1238.

Stubbs, N. and Sikorsky, C. Risk assessment of potential vertical shelters on Galveston Island. *Technical Report 4968 S-6 NSF CEE 83-09511.* College Station, TX: Research Division, College of Architecture, Texas A&M University, August 1987.

Stubbs, N. and Sikorsky, C. Identification of potential vertical shelters in Galveston, Texas. *Technical Report 4968 S-4 NSF CEE 83-09511.* College Station, TX: Research Division, College of Architecture, Texas A&M University, March 1986.

Stubbs, N. and Sikorsky, C. Fault tree analysis as an integrative technique for safety assessment in a wind hazard. *Proceedings of The Fifth U.S. National Conference on Wind Engineering.* K. C. Mehta and R. A. Dillingham, Ed., Texas Tech University, Lubbock, pp. 2B-1-2B-8, 1985.

Tampa Bay Regional Planning Council. Tampa Bay Region hurricane refuse alternative feasibility study. *86 EM-44-12-00-21-004.* St. Petersburg, FL. Department of Community Affairs, Division of Emergency Management, 1986.

Thoft-Christensen, P. and Baker, M. J. *Structural Reliability Theory and Its Application,* Springer-Verlag, Berlin, 1982.

Wallace, J. M. and Hobbs, P. V. *Atmospheric Science An Introductory Survey.* New York: Academic Press, 1977.

Whitman, R. V., Remmer, N. S., and Schumacker, B. Feasibility of regulatory guidelines for earthquake hazards reduction in existing buildings in Northeast. *Publication No. R80-44,* Order No. 687. Cambridge, MA: Department of Civil Engineering, M.I.T., November 1980.

Yao, J. T. P. *An Approach to Damage Assessment of Existing Structures.* Technical Report No. CE-STR-79-4. West Lafayette, IN: School of Civil Engineering, Purdue University, October 1979.

Chapter 4
Hurricane Hazard: An Evaluation of Occupant Safety

Norris Stubbs

Texas A&M University, College Station, TX

4.1 THE EFFECT OF BUILDING TYPE, HURRICANE MAGNITUDE, AND OCCUPANT LOCATION ON OCCUPANT SAFETY

4.1.1 Overview

In this chapter the risk model discussed in the last chapter is used to study the effect of building type, hurricane magnitude, and occupant location on the safety to occupants provided by the structure. For the purposes of clarity, two studies were conducted. In the first study, the safety of occupants in a variety of structures were compared for a range of hurricanes. In the second study, occupant safety as a function of the location of the occupant was investigated.

4.1.2 Description of the Buildings Selected for Study

Six buildings were selected for study here. Each of the structures represented a major class of building to be found on Galveston Island. Plans for the structures were obtained from the appropriate city agency and a visual inspection of each structure and the surrounding terrain was made (Stubbs and Sikorsky, 1986). A detailed description of each of the structures selected is given in the following paragraphs.

Building A. The building is located on the northern side of Galveston Island and functions as a dormitory. Figure 4.1 shows a typical plan and two elevations of the structure. Built in 1983, the building covers a floor plan area of approximately 7,000 square feet, and is two stories high. The height of both stories is 10 feet, and the first floor is 2.5 feet above ground level. This building was assumed to be designed in accordance with the 1982 Standard Building Code.

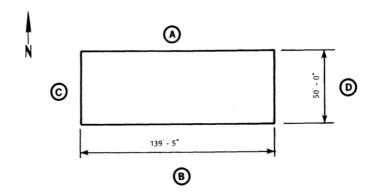

BUILDING PLAN

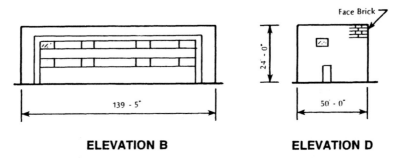

ELEVATION B   **ELEVATION D**

Figure 4.1. Plan and elevation of building A.

A total of seven two-story reinforced-concrete buildings are clustered around the building in question. One hundred feet to the south of this building is a multistory parking garage.

The foundation consists of 6 × 6 feet spread footings, supporting an 8-inch-thick grade beam on the building perimeter, and 7 × 7 feet footings supporting 1 × 1.5-foot floor beams, which span the interior area. Loads are transmitted to the foundation by 1-foot-square concrete columns. The lateral load-resisting system consists of a monolithically poured, moment-resisting frame. The roofing consists of a lightweight aggregate fill covered with 2 inches of insulation and a built-up roof. Wall A faces north, and is a hollow tile infill wall covered with face brick. The remaining three walls are similar in construction. Walls A and B each consist of 20 cladding units, each with dimensions approximately 14 × 10 feet, and walls C and D consist of six units each.

Building B.   The structure is located on the northern side of Galveston Island and functions as a hospital. Figure 4.2 shows two elevations and a typical floor plan of the building. The four-story structure (with a penthouse) was built in 1984, and covers a floor plan area of approximately 45,000 square feet. The height of the first story is 12.6 feet above ground level. Each succeeding floor is 12.4 feet high. The building codes governing the design include the Standard Building Code (1982), the AISC specifications for the Design, Fabrication and Erection of Structural Steel for Buildings (1980), and ACI 318-77.

The surrounding area consists primarily of commercial low-rise buildings. Fifty feet to the north is an unpaved parking lot, and 50 feet to the south is a parking garage, which is connected to the hospital by a walkway. Directly to the west is the original hospital, with which this building shares a party wall. Approximately 300 feet to the east of the hospital is a supermarket.

The foundation consists of grade beams at the building perimeter and spread footings within the interior. A rigid reinforced concrete frame resists the lateral forces. Floors consist of concrete slabs supported by floor beams and joists. The roof is similar to the floor except that the latter also supports a membrane on 5/8-inch-thick insulation board and lightweight aggregate. The exterior walls consist of 4 inch metal studs, with sheeting and face brick on the exterior. Elevation A consists of 28 cladding units, as does Elevation C. There are 36 claddings on Elevation B and 48 on Elevation D. The size of a typical cladding unit is approximately 20 × 12 feet.

Building C.   The structure is located on the southern side of Galveston Island and serves as a hotel. Figure 4.3 shows two elevations and a typical floor plan of the structure. The structure, built in 1983, covers a floor plan area of approximately 16,000 square feet, and has seven stories. The height of the first story is 10.3 feet above

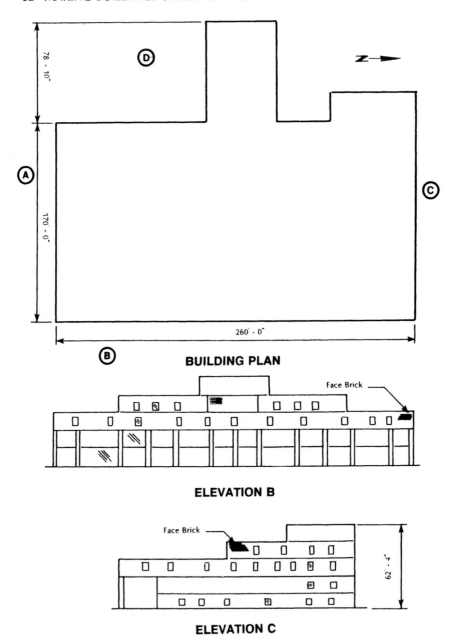

Figure 4.2. Plan and elevation of building B.

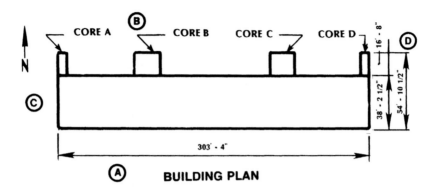

BUILDING PLAN

ELEVATION A

ELEVATION B

Figure 4.3. Plan and elevation of building C.

ground level, and the height of each succeeding floor is 8.7 feet. The height of the seventh floor is 13.4 feet. The building codes governing the design include the Standard Building Code (1979), the AISC Specifications for the Design, Fabrication and Erection of Structural Steel for Buildings (1978), and ACI 318-77.

The terrain to the north, east, and west of the structure consists primarily of low-rise houses and trees. The Gulf of Mexico is to the south of the structure. Fifty feet to the east is a building under construction. Three hundred feet to the north are a row of three-story apartment buildings. One hundred and fifty feet to the west is a 20-story structure. About 200 feet to the south is the seawall.

The foundation consists of 2.25-foot deep reinforced concrete grade beams, supported on 1.33-foot diameter concrete friction piles. The safe load capacity of each pile is specified at 60 tons. The structural material of the superstructure is concrete. Lateral loads are resisted by four cores — two elevator shafts and two stairwells — and thirteen precast, post-tensioned concrete shear walls. The shear walls also support the floors, which consist of precast, prestressed planks. Vertical post-tensioning rods in the cores and shear walls are post-tensioned to 105,000 psi. The shear walls are attached to the grade beam via post-tensioning and grouting, and the floor planks bear on a shoulder in the shear walls and are grouted in place. The roofing support is the same as the other floors, but is covered with a 3-inch-thick insulation board, over which is applied a four-ply roofing membrane. Wall A, which faces the gulf, consists of 168 window wall units. Each unit is comprised of ¼-inch-thick float glass in bronze anodized frames. Walls C and D consist of the end shear walls with a brick veneer. Wall B is made up of a metal stud wall with a stucco finish.

Building D. The structure is located on the southern side of Galveston Island and serves as a banking facility. Figure 4.4 shows two elevations and a typical floor plan of the structure. The structure, built in 1981, covers a floor plan area of approximately 2,500 square feet and is two stories high. The height of the first story is 12 feet above ground level and the height of the second story is 14 feet. The building codes governing the design include the Standard Building Code (1979).

The terrain to the north, east, and west of the structure consists

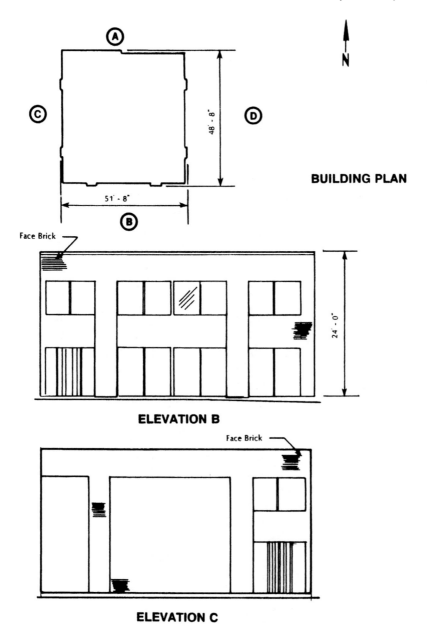

Figure 4.4. Plan and elevation of building D.

primarily of low-rise houses and trees. The Gulf of Mexico lies to the south. To the north of the building is the bank drive-in facility, 50 feet to the east are several two-story houses, and 100 feet to the west is a gas station. Approximately 50 feet to the south is the seawall.

The foundation consists of a grade beam of varying thickness around the perimeter and through the center of the building. The grade beam supports a 5-inch-thick concrete slab, reinforced in both directions. Lateral loads are resisted by a rigid steel frame. Floors and roof consist of 1.33-foot steel joists supporting a steel deck, and 2.5 inches of concrete. The roofing also contains ⅝-inch gravel and a built-up roof over the concrete deck. The exterior walls are constructed of 4-inch metal studs covered with a 4-inch face brick. Each side of the building consists of 8 cladding units. The size of these units is approximately 10 × 12 feet. The window material is ¼-inch tempered glass.

Building E. The structure is located on the northern side of Galveston Island and serves as an academic facility. Figure 4.5 shows two elevations and a typical plan of the structure. The structure, built in 1981, consists of an auditorium and classrooms. Only the classroom portion of the building was considered in the analysis, and the floor plan of that portion is approximately 12,000 square feet and has six stories. The height of the first story is 12.9 feet above ground level, and the height of the second story is 16.9 feet. The next three stories are 13.8 feet high, and the sixth story is 13.1 feet high. The codes governing the design include the Standard Building Code (1979), the AISC Specifications for the Design, Fabrication and Erection of Structural Steel for Buildings (1978) and ACI 318-77.

The terrain on all sides of the building consists primarily of buildings greater than two stories and trees, except for a parking lot to the west. To the north, 200 feet is a six-story academic building. To the south 200 feet is a three-story academic building. Three hundred feet to the east is a three-story library. A parking lot is on the western side of the building.

The foundation consists of a combination of piles and piers. The piles are 76-foot deep, 10.5-inch step-tapered friction type, with a load limit of 40 tons. Piers vary in depth from 64 to 100 feet and are sized accordingly.

Lateral loads are resisted by a rigid steel frame. Both floors and

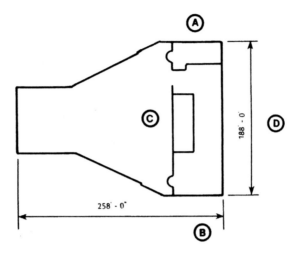

BUILDING PLAN

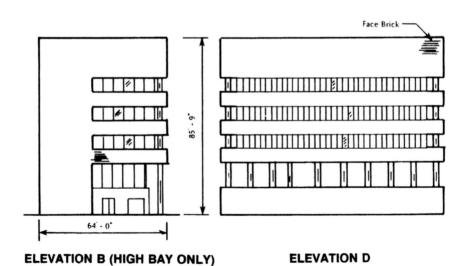

ELEVATION B (HIGH BAY ONLY)   **ELEVATION D**

Figure 4.5. Plan and elevation of building E.

roof are constructed of a metal deck and concrete of varying thickness. This assembly is supported by steel joists, which are supported by the floor beams. Roofing material consists of built-up roofing over lightweight fill and rigid insulation. Exterior walls consist of a 6-inch stud wall system covered with 4-inch face brick. Elevations A and B contain 15 cladding units each. On sides C and D there are 45 units each. The typical size of each cladding unit is 20 × 14 feet. The classroom portion of the building is connected to the auditorium via a glass atrium.

Building F.   The structure is located on the southern side of Galveston Island and serves as a multiple residence. Figure 4.6 shows a plan and two elevations of the building. Built in 1983, the structure covers a floor plan area of approximately 3,100 square feet and has four stories. Each building is divided into four living units, the first floor serving a parking garage. The height of the first story is 11.0 feet above ground level, and the second story is 8.5 feet high. The succeeding two stories are 8.75 feet in height, and the height of the attic is 8.0 feet. The building codes governing the design include the Standard Building Code (1982), the AISC Specifications for the Design, Fabrication and Erection of Structural Steel for Buildings (1977), and ACI 318-77.

The terrain to the north, east, and west of the structure consists of similar buildings. A total of nine buildings of various sizes are within the condominium complex. The greatest distance between any two buildings is 120 feet. To the north, east, and west of this complex are vacant fields. Approximately 50 feet to the south is the seawall and the Gulf of Mexico.

The foundation consists of spread footings six to eight feet square, depending on location. The footings are constructed of reinforced concrete and are two feet thick. The superstructure is comprised of concrete and timber. The living quarters were built on top of the parking structure, which is reinforced concrete. The second level is supported by 2-foot diameter concrete columns resting on the footings. These columns, in turn, support the prestressed concrete beams (1.5 × 3 feet). A precast concrete deck is placed on top of the beams. Exterior walls are of 2 × 4-inch wood studs, spaced 1.33 feet on center. Walls A, B, and D are finished with wood siding on the exterior, and wall C is stucco over metal lath. Roofing is supported

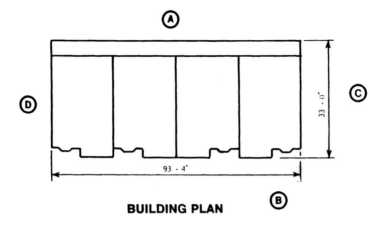

BUILDING PLAN

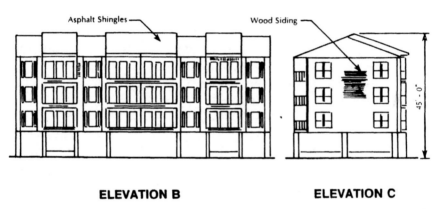

ELEVATION B   **ELEVATION C**

Figure 4.6. Plan and elevation of building F.

by trusses spaced 2 feet on center. Elevations A and B contain 15 cladding units, and C and D contain 3 each. The typical size for these units is 20 × 9 feet.

4.1.3 Method Used to Evaluate the Safety of Occupants

The occupant safety evaluation considered in this section consisted of the following sequence of steps:

1. Plans, specifications, and other pertinent construction documents were assembled for the structure.
2. From the design information and existing information in the literature (e.g., Ellingwood et al., 1980), resistance statistics for the building components were estimated.
3. Given the location and geometry of the structure, the loading statistics for hurricanes of all categories were developed.
4. Failure functions were next defined for the foundation, the lateral load-resisting system, and the roofing and cladding of the structure.
5. Failure functions for each of the building components were evaluated, using methods from Structural Reliability Theory (Thoft-Christensen and Baker, 1982) to yield failure probabilities for the elements, as well as for the major subsystems.
6. The failure probabilities, along with death-damage data (which provided information on the consequences of failure), were used as input for the risk model (Stubbs and Sikorsky, 1987) of the structure to yield estimates of the mean fraction of fatalities and the standard deviation of the estimate. These estimates were provided for each hurricane category.
7. The structure was upgraded to resist a Category 3 hurricane, and Steps 2–6 were repeated.

These steps, and the interrelationships between them, are shown in Figure 4.7. Relevant descriptions of the documents consulted, calculation procedures, loading models, and material properties used in this evaluation are summarized in a later section of this chapter. Detailed computational results for all buildings studied can be found elsewhere (Stubbs and Sikorsky, 1987).

4.1.4 Results of Risk Analysis

The results of the risk analysis are presented in Tables 4.1 to 4.6. Each table lists the risk associated with the structure as a function of the hurricane magnitude and the status of the building (i.e., existing or upgraded). The cost, expressed as a fraction of the replacement cost, to upgrade the structure is also presented. Finally, in interpreting the tables, the reader must keep in mind that the hurricane designations correspond to the Saffir-Simpson scale (see Appendix A).

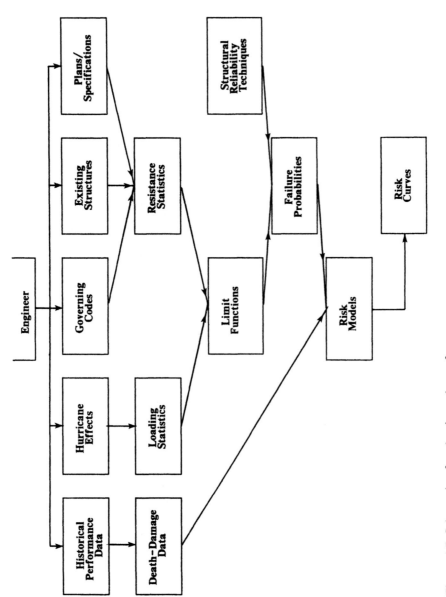

Figure 4.7. Schematic of evaluation scheme for structures.

Table 4.1 Summary of fatality statistics for building A.

| DESCRIPTION OF STRUCTURE | EXPECTED FRACTION OF FATALITIES (STANDARD DEVIATION) | | | | | COST ($000) |
| | HURRICANE CATEGORY | | | | | |
	1	2	3	4	5	
Existing structure ignoring frame and foundation failure	a (a)	.0893 (.2070)	.4210 (.3050)	.4210 (.3050)	.4210 (.3050)	0.0
Structure upgraded[b] to resist Category 3 hurricane	a (a)	a (a)	.0809 (.1990)	.4200 (.3050)	.4220 (.3050)	0.15

[a]Less than 10^{-5}.
[b]Upgrading consisted of the installation of Roll-A-Way shutters, and repairing the roof and the curtain walls.

Table 4.2 Summary of fatality statistics for building B.

| DESCRIPTION OF STRUCTURE | EXPECTED FRACTION OF FATALITIES (STANDARD DEVIATION) | | | | | COST ($000) |
| | HURRICANE CATEGORY | | | | | |
	1	2	3	4	5	
Existing structure ignoring frame and foundation failure	a (a)	a (a)	.0474 (.1570)	.2780 (.2870)	.4220 (.3050)	0.0
Structure upgraded[b] to resist Category 3 hurricane	a (a)	a (a)	a (a)	.0270 (.1210)	.2710 (.2850)	0.31

[a]Less than 10^{-5}.
[b]Upgrading consisted of replacing the window glass and face brick.

Table 4.3 Summary of fatality statistics for building C.

| DESCRIPTION OF STRUCTURE | EXPECTED FRACTION OF FATALITIES (STANDARD DEVIATION) | | | | | COST TO UPGRADE/ REPLACEMENT COST ($000) |
| | HURRICANE CATEGORY | | | | | |
	1	2	3	4	5	
Existing structure ignoring frame and foundation failure	a (a)	a (a)	.3370 (.3260)	.4520 (.3010)	.5410 (.2970)	0.0
Structure upgraded[b] to resist Category 3 hurricane	a (a)	a (a)	a (a)	a (.0004)	.4520 (.3010)	0.23

[a]Less than 10^{-5}
[b]Upgrading consisted of replacing window wall units, exterior stud walls, and strengthening the roof.

Table 4.4 Summary of fatality statistics for building D.

| DESCRIPTION OF STRUCTURE | EXPECTED FRACTION OF FATALITIES (STANDARD DEVIATION) | | | | | COST TO UPGRADE/ REPLACEMENT COST ($000) |
| | HURRICANE CATEGORY | | | | | |
	1	2	3	4	5	
Existing structure ignoring frame and foundation failure	a (a)	.0306 (.1290)	.4210 (.3050)	.4220 (.3050)	.4220 (.3050)	0.0
Structure upgraded[b] to resist Category 3 hurricane	a (a)	a (a)	.0254 (.1180)	.4200 (.3050)	.4200 (.3050)	0.15

[a]Less than 10^{-5}.
[b]Upgrading consisted of the installation of Roll-A-Way shutters and strengthening of the roof and stud wall systems.

Table 4.5 Summary of fatality statistics for building E.

DESCRIPTION OF STRUCTURE	EXPECTED FRACTION OF FATALITIES (STANDARD DEVIATION)					COST TO UPGRADE/ REPLACEMENT COST ($000)
	HURRICANE CATEGORY					
	1	2	3	4	5	
Existing structure ignoring frame and foundation failure	a (a)	a (a)	a (a)	.1050 (.2210)	.2740 (.2860)	0.0
Structure upgraded[b] to resist Category 3 hurricane	a (a)	a (a)	a (a)	a (a)	.0019 (.0339)	0.08

[a] Less than 10^{-5}.
[b] Upgrading consisted of the installation of Roll-A-Way shutters and replacement of cladding with double wythe brick wall.

Table 4.6 Summary of fatality statistics for building F.

DESCRIPTION OF STRUCTURE	EXPECTED FRACTION OF FATALITIES (STANDARD DEVIATION)					COST TO UPGRADE/ REPLACEMENT COST ($000)
	HURRICANE CATEGORY					
	1	2	3	4	5	
Existing structure ignoring frame and foundation failure	a (a)	.0024 (.0368)	.2140 (.2770)	.4160 (.3060)	.4220 (.3050)	0.0
Structure upgraded[b] to resist Category 3 hurricane	a (a)	a (a)	.0033 (.0434)	.2140 (.2760)	.4000 (.3070)	0.07

[a] Less than 10^{-5}.
[b] Upgrading consisted of the installation of Roll-A-Way shutters, additional hurricane clips and joist hangers, and strengthening the stud wall with blocking.

4.1.5 The Effect of Building Type and Hurricane Magnitude on Occupant Safety

If threatened by a hurricane, a potential occupant may desire to know the relative safety of two structures. For our purposes, if two structures are subjected to the same set of environmental conditions, then the structure that exposes the occupant to the lesser risk, however defined, is safer. For example, if the six structures considered here are all subjected to a Category 1 hurricane, the structure with the least risk, as defined in Tables 4.1 to 4.6 is safest. A ranking of the buildings using the criteria above is presented in Table 4.7. The safest building is ranked number 1, and the ranking increases as the risks associated with using the structure increases. If more than one structure has the same risk magnitude associated with it, all such structures are assigned the same ranking. For each structure, the ranking was performed on the basis of the expected fatalities.

In interpreting Table 4.7, the reader should keep two points in mind. Firstly, the comparison is valid only when all structures are subjected to the same hurricane. Thus, for a Category 1 hurricane, the risk associated with using any of the structures is about the same. Secondly, the change in ranking, for a given structure, when the magnitude of the hurricane increases has no meaning. For example, the numbers in rows 1 and 2 have nothing to do with the numbers in rows 3 and 4.

From Table 4.7, several trends are apparent. Firstly, the relative

Table 4.7 Ranking of buildings as they exist.

| | RANKING OF EXISTING BUILDING ON THE BASIS OF EXPECTED FRACTION OF FATALITIES[a] | | | | | |
| | BUILDING | | | | | |
HURRICANE CATEGORY	A	B	C	D	E	F
1	1	1	1	1	1	1
2	3	1	1	2	1	1
3	5	2	4	5	1	3
4	3	2	3	3	1	3
5	2	2	3	2	1	2

[a] 1 = Least risky.

safety of the structure depends upon the magnitude of the hurricane. In a Category 1 hurricane, all structures offer the same level of protection, using the definition of occupant safety used here. Note that this category hurricane is well below the design windspeed for the buildings (the design windspeed for the structure in this group is 110 mph at 30 feet). In a Category 2 hurricane, buildings B, C, E, and F are safest, while buildings D and A are least safe. It is interesting to note that a Category 2 hurricane has windspeeds that are approximately the design windspeeds for the structures. Here we can see that the structures that are greater than 3 stories high are safer than the 2-story structures (A and D). The same trend is repeated for the more extreme hurricanes. Secondly, the variation in the protection offered by the structures as a group is smallest for the extreme hurricanes (Category 1 and Category 5) and largest for the intermediate hurricanes. A similar ranking of the structures is performed for the upgraded structures, and is presented in Table 4.8.

4.1.6 The Effect of Occupant Location on Occupant Safety

So far, our analysis of occupant safety has considered the major elements of the building system (foundation, frame, cladding, and roof) and how the chance of failure of these elements is related to the risk of a fatality. For example, if the exterior cladding on the roof failed, the occupant was assumed to be directly exposed to the fury

Table 4.8 Ranking of upgraded buildings.

HURRICANE CATEGORY	RANKING OF EXISTING BUILDING ON THE BASIS OF EXPECTED FRACTION OF FATALITIES[a]					
	BUILDING					
	A	B	C	D	E	F
1	1	1	1	1	1	1
2	1	1	1	1	1	1
3	3	1	1	2	1	1
4	4	2	1	4	1	3
5	3	2	3	3	1	3

[a]Least risky.

of the hazard. This scenario may be true for a number of buildings (e.g., buildings A, B, and F, which consist primarily of non-load-bearing curtain walls). However, the scenario may not be true for the larger structures. In such cases, the existence of load-bearing walls and massive stairwells present the possibility of a second line of defense. Even if the ordinary cladding or roof fails, the occupants may safely relocate to stairwells or rooms surrounded by reinforced masonry, precast concrete, or other types of load-bearing walls.

To evaluate the impact of utilizing the stairwells when they are present, this subsection summarizes the risk analysis of building C, assuming that the occupants are sheltered in the stairwells of the structure. This analysis will provide at least a ballpark estimate of the extra protection afforded by such structures when they are present.

The structure has two identical seven-story stairwells (A and D, Figure 4.8) located at the western and eastern ends, respectively, of the structure. Each stairwell has interior dimensions of 7.5 × 13.71 feet, and is enclosed by 7-inch-thick post-tensioned precast panels. The total height of the stairwell structure is 52 feet, with stairs connecting floors spaced at approximately 9 feet.

For the purposes of analysis, the stairwell structure is modelled as a cantilever with a constant cross section, shown in Figure 4.9. Failure for the stairwell (i.e., collapse) was defined as follows: 1) The compressive stress at the base of the stairwell exceeds the compres-

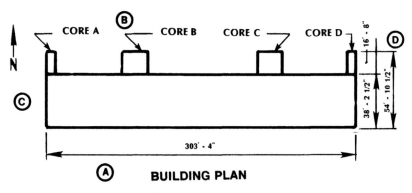

Figure 4.8. Building C stairwell location.

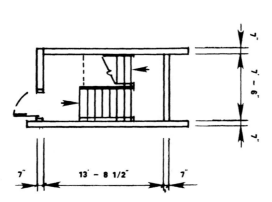

Figure 4.9. Cross section of stairwell structure, building C.

sive strength of the concrete at that section, and 2) at the base of the stairwell, the tensile stress in the prestressing cable exceeds the tensile strength of the cable. Resistance statistics were based on values presented in the plans, and values of the load were determined as discussed in the last section. The failure estimates for the above failure criteria are summarized in Table 4.9.

Note that in Table 4.9, for all hurricane categories, the failure probabilities associated with concrete crushing due to bending only is relatively small, when compared with the failure probabilities associated with the failure of the tendons in tension. As seen in the

Table 4.9 Building core failure probabilities.

HURRICANE CATEGORY	FAILURE DUE TO CRUSHING	TENDON FAILURE	$P(A \cup B)^b$	REVISED P_F (ONE CORE)
1	a	3.362×10^{-3}	3.362×10^{-3}	a
2	a	6.789×10^{-3}	6.789×10^{-3}	a
3	a	7.373×10^{-3}	7.373×10^{-3}	a
4	a	8.376×10^{-3}	8.376×10^{-3}	8.376×10^{-3}
5	a	9.626×10^{-3}	9.626×10^{-3}	9.626×10^{-3}

[a] Less than 10^{-6}.
[b] A = {Failure due to crushing}; B = {Tendon failure}.

next column, the failure of the post-tensioning system controls the safety of the structure. In the last column, an *a posteriori* probability of failure of the core is estimated on the basis of the observed behavior of engineered structures in a wind environment. For a Category 1, 2, or 3 storm, no engineered structure (in the U.S.A.), subjected only to wind, has been observed to collapse. Therefore, if we define the event $\{E\}$ to be "the occurrence of a Category 1, 2, or 3 hurricane" and $\{F\}$ to be the event that "the structural frame collapses", we may write

$$P[F|E] = P[F \cap E]/P[E] \qquad 4.1$$

But, so far, the event $\{F \cap E\} = \{\phi\}$, where $\{\phi\}$ is the null set. Therefore, since $P[E] \neq O$, $P[F|E] = O$. Note that this latter probability estimate is based on past experience, and there is no guarantee that, in the future, the event $\{E \cap F\}$ will not occur. The revised failure probabilities for one core are listed in the last column of Table 4.9. Because of a lack of knowledge of the behavior of engineered frames in Category 4 and 5 hurricanes, no updated estimates of the probabilities are proposed here.

To complete the risk calculation, the following modifications were made to the original calculations: 1) instead of neglecting the frame failure, the probability of failure of either stairwell A or D was assumed; 2) since it was assumed that the inhabitants would all be sheltered in the stairwells, the exposure in the event of roof, cladding, or partition failure was set equal to zero. The resulting input data are given in Table 4.10. The expected fraction of fatalities and the associated standard deviations for each hurricane category are listed in Table 4.11. The measures of risk expressed in terms of the expected number of fatalities per 100 persons, and the various probabilistic descriptions, are listed in Table 4.12.

Several interesting observations can now be made, if we compare Tables 4.11 and 4.12 with Table 4.3. Without introducing any modifications to the existing structure, but by simply modifying the exposure of the occupants (i.e., locating them in different parts of the building), fatalities are dramatically reduced. This observation is particularly true for Category 3, 4, and 5 hurricanes. Put another way, the use of the stairwells dramatically increases the chances of occupant survival in the more intense hurricanes. For example, in a

Table 4.10 Risk model data input for building C, assuming all evacuees are in stairwells.

		BASIC EVENT PROBABILITIES				
		HURRICANE CATEGORY				
BASIC EVENT	DESCRIPTION OF BASIC EVENT	1	2	3	4	5
X_1	Hurricane occurs	1.00	1.00	1.00	1.00	1.00
X_2	Lateral forces exceed frame strength	a	a	a	1.67×10^{-2}	1.93×10^{-2}
X_3	Person is exposed|frame fails	1.00	1.00	1.00	1.00	1.00
X_5	Foundation fails	a	a	a	a	a
X_6	Person is exposed|foundation fails	1.00	1.00	1.00	1.00	1.00
X_8	Uplift forces exceed roof strength	a	a	0.74	1.00	1.00
X_9	Person is exposed|roof fails	0.00	0.00	0.00	0.00	0.00
X_{11}	Lateral forces exceed cladding resistance	a	a	a	a	1.00
X_{12}	Person is exposed|cladding fails	0.00	0.00	0.00	0.00	0.00
X_{14}	Lateral forces exceed int. part. resistance	a	a	a	a	a
X_{15}	Person is exposed|partition fails	0.00	0.00	0.00	0.00	0.00

ªLess than 10^{-7} or failure mode ignored.

Table 4.11 Risk of using the stairwells of existing building C in various hurricanes.

HURRICANE CATEGORY	EXPECTED FRACTION OF FATALITIES	STANDARD DEVIATION OF EXPECTED FATALITIES
1	a	5.46×10^{-5}
2	a	5.46×10^{-5}
3	a	5.46×10^{-5}
4	4.56×10^{-3}	5.08×10^{-2}
5	5.26×10^{-3}	5.45×10^{-2}

ªMagnitude $< 10^{-7}$.

Table 4.12 Risk of using stairwells of existing building C.

MEASURE OF RISK	MAGNITUDE OF RISK				
	HURRICANE CATEGORY				
	1	2	3	4	5
Expected Fatalities[a]	0	0	0	0	1
Someone is killed: P($X > 0.05$)	0.000	0.000	0.000	0.014	0.017
No one is killed P($X < 0.05$)	1.000	1.000	1.000	0.986	0.983
50% or more are killed P($X > 0.50$)	0.000	0.000	0.000	0.001	0.001
All are killed P($X > 0.95$)	0.000	0.000	0.000	0.000	0.000

[a]Number of fatalities per 100 persons.

Category 4 hurricane, the probability that someone is killed if the stairwells are used exclusively is reduced from unity (i.e., certainty) to less than 2 in 100. Simultaneously, at the other extreme, the probability that 50 percent or more are killed has been reduced from 1 in 2 (for the case of not using the stairwells) to 1 in 1000 (for the case of using the stairwells). Finally, the option of using stairwells (particularly in Category 3, 4, and 5 hurricanes) is superior to the option of upgrading the structure. In conclusion, we make the following point: Although people are aware of the additional protection offered by load-bearing walls or stairwells, we have presented here one of the first quantitative estimates of such added protection.

4.2   SUMMARY AND CONCLUSIONS

This study has considered the question of occupant safety from the viewpoint of the occupant of the structure. The expected fraction of fatalities, in a structure subjected to a hurricane, is used here as the measure of risk. The lower this value (i.e., the expected fraction of fatalities) for a given structure, the safer the structure is, from the occupant's perspective. We have shown how this risk can be rationally expressed as a function of the structural and nonstructural characteristics of the building, the exposure of the occupant, and the consequences of failure of the structural and nonstructural subelements. We have also shown how data needed to express the behavior of the various components of the building can be obtained from existing building documents, the characteristics of the hurricane, and data published in the literature. These data have been used to produce a quantitative estimate of occupant safety via a specification of the first two moments of the random variable: the fraction of occupants in a structure that will die if the structure receives a direct hit from a hurricane.

We have shown, on the basis of a detailed analysis of a variety of existing buildings, how occupant safety varies with building type. In the sample of buildings studied here, multistory structures greater than three stories—whether they be steel or concrete—are safer than buildings of two stories or less. We have also shown, quantitatively, how the location of the occupants of the structure affects the chances of survival of the occupant. Seeking shelter in stairwells or other specially protected areas increases the chance of survival by orders of magnitude. Finally, we have shown that the protection offered by a structure can be increased significantly by modifying the nonstructural elements, such as roofs and exterior cladding.

Although these findings are generally known in a qualitative sense, the procedure presented here provides a quantitative assessment of occupant safety provided by a structure. The advantages of such an approach are twofold. Firstly, the focus here is on the safety of the occupant as part of the total building system and not, as is conventionally assumed, as a single function of the safety of the structural portion of the building system. Secondly, this technique can be used in more general problems, in which mitigative options other than structural options are to be evaluated. If, for example, administrators in a coastally located hospital wanted to decide between the option of keeping critically ill patients in the hospital, subjected to the hurricane, or the option of transporting the patients elsewhere, the bottom line is the comparison between the expected number of fatalities in the two cases.

In conclusion, although the model, *per se*, is comprehensive, much research is needed to improve the estimates of the component risks. Future work is needed to express the collapse state for frames, foundations, cladding systems, and roof systems. Much work is also needed to define the consequences of collapse of the various structural subsystems.

ACKNOWLEDGMENTS

The financial support of the National Science Foundation, under Grant CEE 83-09511 to Texas A&M University, is gratefully acknowledged. The assistance of Ms. Patricia Lombard in typesetting the manuscript and helping with the proofreading is also greatly appreciated. Finally, the contribution of Mr. Charles Sikorsky, P.E., who performed the structural inspections and the cost analysis, and who contributed significantly to the risk analysis, is also greatly appreciated.

REFERENCES

American National Standard, ANSI, 1982. *Building Code Requirements for Minimum Design Loads in Buildings and Other Structures, ANSI A58.1.* New York: American National Standards Institute, 1982.

AISC, *Manual of Steel Construction, 1970.* New York, NY: AISC, 1970.

American Concrete Institute, *ACI Manual of Concrete Practice, 1977.* Detroit, MI: ACI, 1977.

AISC, *Manual of Steel Construction, 1978.* New York, NY: AISC, 1978.

Ellingwood, B., Galambos, T. V., MacGregor, J. G., and Cornell, C. A. Development of a probability based load criterion for American National Standard A58: Building Code Requirements for Minimum Design Loads in Buildings and Other Structures. *NBS Special Publication 577.* Washington, D.C.: U.S. Department of Commerce, June 1980.

Southern Building Code Congress International, Inc. *Standard Building Code, 1979.* Birmingham, AL: Southern Building Code Congress International, Inc., 1979.

Southern Building Code Congress International, Inc. *Standard Building Code, 1982.* Birmingham, AL: Southern Building Code Congress International, Inc., 1982.

Stubbs, N. and Sikorsky, C. Risk assessment of potential vertical shelters on Galveston Island. *Technical Report 4968 S-6 NSF CEE 83-09511.* College Station, TX: Research Division, College of Architecture, Texas A&M University, August 1987.

Stubbs, N. and Sikorsky, C. Identification of potential vertical shelters in Galveston, Texas. *Technical Report 4968 S-4 NSF CEE 83-09511.* College Station, TX: Research Division, College of Architecture, Texas A&M University, March 1986.

Thoft-Christensen, P. and Baker, M. J. *Structural Reliability Theory and Its Application.* Berlin: Springer-Verlag, 1982.

Chapter 5
Fire Hazard: Historical Perspectives

Paul R. De Cicco, P.E.
Polytechnic University

5.1 THE SAFETY OF OCCUPANTS FROM FIRE

5.1.1 Perspectives

From the earliest times of man's construction of dwellings, as well as other shelters to protect food supplies and other materials he wished to store, there was recognition that such structures needed to be safe from water and winds, from the effects of the sun's heat and even from the movement of the earth on which they stood. Protection from fire was not regarded in quite the same way, perhaps because (unlike other perils) fire became man's friend early on, as he brought it inside to keep warm, to cook, and to heat water.

In some respects, this situation has not changed much over the millennia since man emerged from his first cave habitat. In a strict sense, neither the engineering profession nor the profession of architecture regard optimization of buildings to resist fire as a primary design consideration, or in a holistic manner. In fact, many of our buildings, particularly small residential structures (which are the most numerous and in which most fire fatalities and injuries occur), are not engineered at all.

Good wind design practice requires that buildings be safe against overturning and sliding, that they do not vibrate at dangerous frequencies or deflect beyond specified amounts. Other dead and live loads are meticulously calculated, so that structural collapse under a wide range of loading conditions that can be anticipated is indeed a rare event. However, there is no equivalent mandate that, in a building attacked by fire, occupants will (with any certainty) be able to exit safely within the time available or that they will somehow be able to remain safely within the structure.

Instead, style, configuration, and details that satisfy the architect's creative inclinations and the uses that the building will serve are set first. Following this, mechanical and electrical systems needed to provide light, conditioning of air, vertical transportation, and other utilities are introduced. Only then are fire protection measures added, made to fit functional needs and verified against minimum building code requirements.

While this is a somewhat simplistic view, its essence is that even the most essential requirements for life safety from fire, such as safe means of egress, compartment sizes that will assure a controllable fire anywhere in the building, and provisions for the control of smoke (which is responsible for most fire deaths) are not optimally designed with fail-safe features, but rather to serve functional and economic demands.

For example, we don't use hanger-sized fire exit doors, even in spaces that accommodate thousands of people, despite the hard lessons learned from such tragedies as the Coconut Grove Nightclub, Beverly Hills Supper Club, and Dupont Plaza Hotel and Casino fires. Designers are also more inclined to provide as few stairs as codes will permit, rather than a number which will assure escape from the ends, center, or any other location in a specific building configuration.

It is indeed a sad exercise to examine the circumstances in some of the worst building fires, with respect to loss of life, and to note that only a single additional exit in an obvious location would have allowed many, if not all, occupants to escape safely. When New York City promulgated its current building code in the late 1960s, it abandoned a longstanding requirement for smoke-proof stairs, in the interest of more economical construction. It took most of the following decade for the same city to pass new legislation with costly, retroactive requirements regarding fire stair safety.

Fire safety, rather than a primary and universal goal, assumes more of a passive role in the design and engineering of buildings. Materials are permitted in buildings as finishes and furnishings that are known to be hazardous, with respect to their ignitability, the quantity of energy they will release, the rate at which they will release energy, and the toxicity of products of combustion that they emit. Fire protection measures are then added to mitigate the hazards thus created.

Perhaps the reason why, in spite of all of the impressive technological advances made by the U.S. building industry during this century, we have not reduced our tragic fire statistics in a commensurate way, is because fire safety construction features and benefits are not permitted to accumulate. Instead, each time a new material, method, or device that can enhance fire safety emerges, there is a tendency to remove or reduce some other measure that is already in place.

For example, it is only recently that bickering has ended in the fire protection community regarding whether both detectors and sprinkler systems are really necessary in high-rise buildings. Debate continues, however, over the simultaneous use of sprinklers and smoke removal systems, and there is a constant thrust from manufacturers and builders to reduce the fire resistance of walls in buildings that have sprinkler systems.

In recent years, regulatory authorities have begun to follow policies that embrace, rather than shun, the use of what were referred to as redundant fire systems, and which encourage the simultaneous use of fire safety measures, including comprehensive early warning systems, truly complete sprinklerization, controlled compartment sizes, generous numbers of protected fire stairs, and control of combustible finishes and furnishings. New attention is also being directed to the requirement of many of these measures in one- and two-family residential buildings, where most fire casualties occur.

In the meantime, the engineer should resist the temptation to address the safety of occupants solely by following minimum requirements set forth in building codes, especially when buildings have unusual features for which there is little or no fire experience. The engineer must also see to it that fire safety measures specified compensate for the absence of basic safety attributes such as the use of noncombustible materials for linings and other interior uses, true redundancy of exits, and furnishings with reasonable fire properties.

The engineer should also, by conservative design, acknowledge the realities of poor workmanship, less than adequate code enforcement, poor maintenance, and possibilities for less than rational human behavior during fire emergencies.

5.1.2 Scope and Focus

This chapter considers the responsibilities of the building design professions with regard to the performance of buildings exposed to fire. The primary focus is directed to those qualities that affect the life safety of occupants. Fire experience and recent fire statistics are reviewed, fire phenomena are described, and properties and performance characteristics of common construction materials are discussed. Important fire safety instruments, such as the fire safety plan, the fire protection plan, and building codes and standards, are discussed. The effects of unwanted fire on the structure, its occupants, and fire fighters are also considered.

Means and measures for mitigating the consequences of building fires, and the hazards they present to occupants, are presented. Issues involving life safety in building fires are discussed in the context of current design practices and trends in building architecture, which feature larger, taller, more complex structures, housing greater numbers of occupants, and which support increasingly diverse activities.

For more definitive information and computational models for executing actual designs of components and systems alluded to herein, the reader is referred to specific building and fire codes, fire protection design manuals and a wide range of technical literature available from a number of sources.

These sources include the National Fire Protection Association, the Society of Fire Protection Engineers, the Center for Fire Research of the National Bureau of Standards, and a large number of testing laboratories, manufacturers, industrial associations, and fire research organizations. Their respective publications provide a wide range of information, instruction, and models for analyzing fire phenomena, and for executing designs of building components and systems.

The reader is reminded that adequate protection of a building does not necessarily assure satisfactory levels of life safety for occupants and, conversely, providing high levels of life safety for occupants does not assure protection of the building under severe fire conditions. The two issues must be addressed separately, but with sensitivity for their many interrelationships.

5.2 FIRE EXPERIENCE IN THE UNITED STATES

Throughout almost all U.S. fire history in this century, and continuing to the present, there have been many tragic demonstrations of consequences that have derived from ignoring well-established tenets of fire safety in buildings. Perhaps the greatest frustration of fire safety specialists and members of the fire services is the fact that it is extremely rare to find a fire of serious human consequence that could not have been avoided if lessons already learned had been heeded.

5.2.1 Historical Fires and Current Implications

One of the earliest and perhaps the most disastrous fire event of this century occurred in the Iroquois Theatre in Chicago in 1903 (Backes, 1973). This fire killed 602 persons, the largest number of fatalities in a threatre fire ever to occur in the United States. It was caused by a stage arc light that ignited a large combustible stage curtain. Unmarked and obstructed exits were blamed for the high loss of life in this "newest and safest" of theatres.

It is of some interest to note that, in the late 1980s, our "newest and safest" theatres are being built many stories above the street level in high-rise buildings of considerable complexity, and by no means free of fire risk. In addition, traditional separation of stage areas and audience seating is disappearing as more stages are extended into audience space. Older houses are also being similarly renovated to "bring performers into the audience." Recent stagings of "Cats" and "Starlight Express" in New York City are examples of this trend.

In the former production, combustible stage materials and props simulating a refuse dump were carried into audience seating areas of both orchestra and upper levels. In the latter show, a large stage area was extended into the audience to accommodate a raised wood-floor roller skating rink.

These kinds of innovations, while serving the best traditions of the theatre, breach time-honored principles that have long favored the location of places of large public assembly in immediate and direct communication with street levels, providing exterior exits no more than two or three stories in height, and keeping stage areas, with their special lighting, electrical machinery, and combustible material hazards, quite separate from audience seating areas.

Such undertakings present fire departments with serious new concerns. While "on-the-spot" corrective measures, such as the installation of improvised sprinkler systems and the hasty development of emergency exit paths are called on as "one-time fixes," these new trends certainly represent new hazards to occupants. Recent interest in finding new ways to use existing buildings, and the difficulties associated with patchwork efforts to fit reliable fire safety measures to them, may suggest why (over the long run) fire statistics in this country have not improved as much as might be wished for and expected.

In 1911, the famous Triangle Shirtwaist Company fire in New York City resulted in 145 employee deaths, and here again, inadequate (as well as locked) exits were the chief impediments to escape. The building, which cost $400,000 and was opened in 1901, was said to be "fireproof," but had wood floors and wood window sashes, and it was decided for economic reasons not to install a sprinkler system, which would have cost $5000 (Backes, 1973).

Poorly marked exits are still common hazards in many public buildings, and in these times of great concern for building security, it is not uncommon to find doors from public assembly areas chained or otherwise illegally locked, even in times of full occupancy.

The locking of fire stair doors is also increasingly practiced in the interest of building security. It then becomes critical to assure that these doors will be automatically and promptly opened during fire emergencies.

The Ohio State Penitentiary fire (Backes, 1973), in 1930, killed 320 persons, and during the past 20 years, there have been dozens of additional fires causing multiple deaths in correctional institutions in the United States. Even as this is being written, New York City has undertaken a project that will place prisoners in floating barges as "temporary" holding facilities, adding new and significant dimensions to the task of providing adequate means for evacuation in time of fire.

Four hundred and fifty-five persons perished in the Consolidated School explosion and fire in New London, Texas, in 1937. This event was caused by natural gas leaking into the basement of the school. The gas was said to have entered a large, and poorly ventilated, concealed space containing gas pipes and electrical conduits. The ignition source was declared to be a spark from an electric switch. The building was of steel frame construction with wall columns supporting the roof trusses (Ross, 1984).

Gas explosions and fires resulting from improper use or lack of safety devices associated with the combustion of fuels in buildings are still among the most feared and most destructive events, with respect to occupant safety.

Problems associated with concealed spaces in buildings continue to be among the most serious of concerns in terms of fire hazards created. Closely associated are wall and floor penetrations that are not properly fire-stopped. The extension of fire through such spaces and openings was responsible for the scale of damage produced in the One New York Plaza fire in 1970.

Three floors of this 50-story office building were involved in a body of fire estimated to cover over 48,000 square feet in area. It took over 5 hours for fire fighters to control the fire, and 191 steel beams had to be replaced or strengthened as a result of failure of the fireproofing, and 21,000 square feet of concrete floor also had to be reconstructed. Over $10 million in damage was reported. Two employees were killed and there were 30 injuries (O'Hagan, 1977).

This fire, and one in a 47-story building at 919 Third Avenue in New York City in the same year (which killed 3 persons and injured 20) were largely responsible for the promulgation, in 1973, of New York City's Local Law Number 5. This law (which has undergone a number of revisions over the past decade) was directed to the enhancement of fire safety of occupants in high-rise buildings, and retroactively regulated commercial buildings over 100 feet in height.

Among its requirements were: the development of a fire safety plan, installation of a fire command station, smoke detectors, alarms, and two-way communication systems, limits on compartment sizes in unsprinklered buildings, controlled use of elevators, options for protection of fire stairs by pressurization means, and new specifications for fire emergency signs.

In 1942, the notorious Coconut Grove Nightclub fire caused the deaths of 492 people. Highly combustible finishes, furnishings, and decorations were primary factors in the rapid spread of the fire. The nightclub held over 1000 patrons, although it was licensed for only 460. Walls and seating along walls were covered with combustible imitation leather. Ceilings were decorated with streaming lengths of cloth, and papier-mâché palm trees and lacquered paper foliage were placed on and near wood tables throughout the room of fire origin. Contributing to the high loss of life were the facts that exits were not marked, people tried to leave by way of constraining revolving doors through which they had entered, and general panic occurred (Backes, 1973). Modern codes address the quantities and requirements for flame-retardant treatment of decorations in public places.

5.2.2 More Recent Fires of Note

In 1958, a fire in Our Lady of Angels School in Chicago left 94 dead, almost all of whom were children. The building was constructed of combustible materials, was unsprinklered, and included open stairs. There was an early collapse of the timber-supported roof and of walls and partitions, which trapped victims in classrooms. These events made this fire unusual, as building collapse as a cause for casualties is rare (Backes, 1973).

In more recent times, and to a large extent providing notice of the persistence of the same types of deficiencies observed in earlier tragic fire incidents, there have been a large number of deadly and costly fires in hotels, motels, and places of entertainment.

In a fire of electrical origin, intricate routes of egress were held to blame for the loss of 165 lives in the Beverly Hills Supper Club in Southgate, Kentucky in 1977 (Best, 1978 and Emmons, 1983). This fire occurred at approximately 8:45 PM. Occupants of the crowded Cabaret Room were not made aware that a fire was in progress for over 20 minutes. Major deficiencies regarding good practice in the protection of occupants included the use of unprotected plywood walls, a lack of alarms and sprinkler systems, and exits insufficient in size and inadequate in number for the large assembly of people.

It is estimated that the fire traveled from the space of origin, down a 150-foot corridor to the dining room in which most of the fatalities occurred, in less than 5 minutes. It is believed that the fire was burning for some time in a concealed ceiling space before discovery. The Cabaret Room, like a number of the large gambling casino spaces that have been so prominent in recent disastrous fires, was an extremely large, uncompartmented space estimated to be almost 230,000 square feet in area.

While fatalities were due almost entirely to smoke and carbon monoxide inhalation, and not to structural failure, many of the circumstances involving the initial fire development and movement of fire and smoke from the room of origin to the Cabaret Room, and the inability of occupants to safely evacuate, resulted from poor basic building design.

The large uncompartmented cabaret space in an unsprinklered building, the absence of effective smoke and fire barrier doors in connecting corridors, the presence of communicating concealed spaces, the undersized and limited number of exits available to patrons of the main room of assembly, and the absence of any kind of engineered smoke control system, may all have been allowed by the building code in force, but certainly were fire safety deficiencies that should have been avoided.

The MGM Grand Hotel fire in Las Vegas in 1980 killed 85 people, and injured over 500 (Breen, 1981). This fire (also of electrical origin), in a partially sprinklered building, moved rapidly through concealed spaces and caused over $50 million in damages and litigation in the hundreds of millions of dollars.

It is reported that the open, 200,000 square-foot casino area was destroyed in ten minutes, and that smoke penetrated many of the upper floors and was also distributed through duct systems. Sixty-one victims died on the sixteenth through twenty-sixth floors.

This building must be considered a modern structure, having been constructed in 1973. It is of some interest to note that the claim that no sprinklers were necessary in the casino, because it operated 24-hours per day, and therefore fire would be immediately observed (which is the same claim made for large atrium spaces in hotels), is not always valid.

Extensive violations of local building codes were cited following this fire and included inadequate exit widths, improper signs and penetrations of air shafts, and large openings in corridor walls. The consequences of this fire must be ascribed to both code deficiencies and inadequate enforcement of such code provisions as were in force.

Other notable fires in hotels occurred in the Las Vegas Hilton in 1981, in which there were 8 fatalities and 350 injuries, and in the Stouffer's Inn at Harrison, New York in 1980 (Bell, 1982). In the latter incident, 26 died and 24 were injured in the Inn's conference facility, as a result of a fast arson fire set in a corridor. Some of the meeting rooms in which occupants were trapped had no remote exits and the building was unsprinklered.

Major factors contributing to the loss of life were given in a National Fire Protection Association report. They included critical location of the fire at the intersection of exit access corridors, rapid development of the fire, the lack of remote, second means of egress from meeting rooms, and the lack of a fire protection system to detect and extinguish the fire at its incipient stage.

The Westchase Hilton Hotel in Houston, Texas, which opened in 1980, was the scene of a fire that killed 12 people in 1982. Stairs were located only at the ends of corridors, which rapidly became smoke logged and prevented victims from reaching them.

5.2.3 Dupont Plaza Hotel and Casino

In many ways, the fire that occurred in the Dupont Plaza Hotel and Casino in San Juan, Puerto Rico on December 31, 1986, serves as a model for "lessons not learned" with regard to providing for the safety of occupants in buildings. This fire caused 98 fatalities and hundreds of injuries. It was an arson fire, but many investigators believe that most of the dire consequences would have been the same had the fire been ignited in the same location by accident.

The fire originated in stored furniture and associated packing materials in a ballroom, and quickly spread through lobby areas and into the Casino. The building was unsprinklered and exits proved to be insufficient under the fire conditions that existed. An important circumstance in the failure of many to escape the casino was the delay of staff in advising patrons of the fire emergency and of the need for immediate evacuation. The fire occurred at approximately 3:30 PM, when staff and patrons were awake and alert, and theoretically in the best position to respond to the emergency.

The National Bureau of Standards, in April of 1987, issued a report entitled *An Engineering Analysis of the Early stages of Fire Development — The Fire at the Dupont Plaza Hotel and Casino*, which addresses the development and growth of the fire up to the time of the emergence of flame from the Casino (National Bureau of Standards, 1987).

This document illustrates the kinds of information needed and the types of computations that can be made in post-fire analysis. Some of the computations included in the report are: mass burning rates during pre-flashover burning; post-flashover burning; rates of heat release and energy flows between spaces; smoke temperatures in the various spaces; smoke layer depths and velocity of smoke front; viability in smoke layer; flame spread; and potential responses of sprinklers and smoke detectors.

The analyses presented used fire growth models, engineering formulae, and technical data. They provide an excellent basis for understanding much of the fire dynamics involved in large fires in general, and in this fire in particular. This report should be of great interest and value to building designers and others concerned with the enhancement of occupant safety in buildings.

The analysis begins with an assumed initial fire size, with an energy emission of 5 to 10 kw (kilowatts). A number of findings from this study are of interest here, and are outlined below. They illustrate obvious and gross digressions from good practice with regard to the life safety of occupants.

- The initial fire load was a volume of approximately 3000 cubic feet of stacked wood and particle board furniture in corrugated cartons located in a ballroom. The estimated mass of this initial fuel load was between 30,000 and 45,000 pounds.
- Ignition was by means of accelerants placed in or on the stored furniture, and therefore was both direct and rapid. The room was unoccupied.
- Overall, heat of combustion of fuels consumed was considered to average approximately 12,000 Btu/lb, with appreciable amounts of wood and plastic materials present.
- Two walls of the room of fire origin contained fabric coverings, and a dividing wall was partially open.
- Unrated glass partitions separated the room of fire origin from an adjacent foyer area that had wood ceilings and into which the fire spread.
- Ceilings in the path of fire travel were of mineral board and wood, doors and floors were of wood, and ceiling spaces offered paths for fire extension.
- The building was unsprinklered.

These conditions, each conducive to rapid fire growth and extension, resulted in a number of observed and calculated manifestations, including:

- Flashover in the South Ballroom (room of fire origin) occurred in 10 minutes.
- Forty seconds after flashover, exits from the casino were blocked by flame and high carbon monoxide levels.
- Occupants were not aware of the emergency for ten minutes, which is approximately the time fire breaks out of the room of origin and into the foyer.
- Fire swept through the casino in approximately 20 seconds, at velocities of 6 feet per second, giving little chance for escape through limited available exits.
- Energy release rates from the South Ballroom to foyer were on the order of 89,000 BTU/sec (94 MW [megawatts]).
- Energy release rates from the foyer ceiling were on the order of 104,000 BTU/sec (110 mw).

This account shows how unfavorable and hazardous conditions that existed in the building resulted in a fire with extremely devastating effects on occupants. Fuel loads were excessively high, walls between the South Ballroom and adjacent foyer were not sufficiently resistant, materials used for finishes on walls and ceilings permitted rapid flame spread, warning systems and building staff actions were deficient, and occupants, even after becoming aware of the existence of a fire, failed to leave the casino in a timely fashion.

The high energy release rates, fast flame velocities, and extreme temperatures reached during the first few minutes following ignition suggest the severity of exposure that structural components may experience within the first 10 minutes or so of a fast fire. These substantial rates of energy release and other reported conditions, including descriptions of a wall of fire moving to and through the casino, should remind designers that laboratory test environments may not bear much relationship to actual fire conditions, and that they should take time to become acquainted with the realities of large fires in buildings.

5.3   RECENT FIRE STATISTICS

In addition to consideration of specific case histories of serious fire occurrences, building design professionals and others interested in the enhancement of fire safety for occupants can learn much from fire statistics. Familiarity with the frequency of fires in various occupancies and types of construction, knowledge of locations within buildings where fires are most likely to originate, and common causes of ignition and manner and rate of fire growth, can all be of considerable value in anticipating the nature of fire exposure to which building components may be exposed.

A quantitative perspective of the consequences of building fires, in terms of civilian and fire fighter casualties, and in the context of the nature and extent of property damage and other losses incurred, may be gained through the examination of fire records. Because economic considerations play so prominent a role in the decision-making process associated with building construction, statistics that portray losses from fire should be constantly reviewed by all components of the building community.

5.4 SOURCES OF FIRE DATA AND A DISCUSSION OF CASUALTIES AND DAMAGES

There are three national data bases available that report fire experience in the United States. They include the Annual Survey of Fire Departments, by the National Fire Protection Association; the National Fire Incident Reporting System (NFIRS) of the United States Fire Administration (FEMA/USFA); and the NFPA Fire Incident Data Organization.

These three data bases are organized differently, and vary in the detail of fire information contained. However, together they permit reliable estimates and projections to be made of where and how fires are likely to occur in buildings. To a more limited extent, they also permit examination of the physical circumstances leading to casualties and property damage.

5.4.1 Fire Deaths

The National Fire Protection Association reports that, from 1977 to 1984, an average of 6633 civilians died each year from fire in the United States. Approximately 85 percent of these occurred in structural fires, and approximately 96 percent of structural fire deaths occurred in fires in residential occupancies.

Multiple-death fires in the U.S. (fires in which three or more deaths occur) number approximately 1000 annually, which is more than the combined total from earthquakes, tornadoes, hurricanes, floods, and other natural disasters.

During the period from 1977 to 1984, residential occupancies accounted for most (65 percent) of both incidents and deaths in multiple-death fires resulting in ten or more deaths (National Fire Protection Association, 1984).

While there has been a decline in annual deaths from fire over this period of record, current statistics still place the United States at the forefront of all industrialized countries on the basis of deaths per unit of population or per given number of buildings.

5.4.2 Occupant Injuries

It is estimated that there are in excess of 130,000 civilian injuries due to fires each year. Here again, the largest number of these (69 percent) are attributed to residential fires. Of all fire statistics, those relating to injuries are perhaps the poorest reported. It is universally agreed in the fire safety community that the actual number of injuries sustained from fire each year by civilians may be greater than reported by an order of magnitude.

In considering reasons for the severity of civilian casualty statistics in the United States, one must be aware of the fact that the number of reported fire incidents averaged over 2.7 million annually, between 1977 and 1984, with the number of structural fires numbering approximately 994,000, or 36 percent, of all fires.

The extremely large and diverse use of electrical energy in industry (here, approximately 25 percent of all fires are attributed to electrical causes), in commercial buildings and in the home, as well as a number of behavioral factors attributed to the U.S. population, have been cited as contributing to these statistics.

Some important conclusions that may be derived from fire statistics in the United States are that:

- Most fires occur in buildings.
- Most deaths and injuries from fire occur in buildings.
- Most fire casualties occur in residential buildings.
- Most multiple-death fires occur in residential buildings.
- Most fire deaths and injuries result from smoke and gases.

It is also of interest to note that only one (heating and equipment) of the four leading classes of ignition that cause deaths in buildings (which are: smoking materials, incendiary and suspicious, heating equipment, and cooking equipment), relates directly to the responsibilities of building engineers.

The type of structure that accounts for the majority of fire deaths (80 percent), injuries (67 percent), and property damage (61 percent), is the home (1- and 2-family dwellings, mobile homes, and apartments). This type of building is almost entirely "unengineered," which indicates that, while most engineering efforts in the building industry are associated with commercial, industrial, and public buildings, these are not the environments which contribute significantly to the nation's dismal fire statistics, regarding deaths and injuries.

Figure 5.1 shows fire tests conducted in a "high-risk" residential neighborhood. Wood frame buildings such as those shown produce fast and deadly fires. Figure 5.2 depicts the aftermath of a fast fire that involved sixteen buildings in a single incident.

Recent efforts directed toward the improvement of life safety from fire in residential buildings have focussed on after-the-fact defensive measures. Thus, most attention has been given to the installation of smoke and fire detectors and sprinkler systems, rather than to basic elements of the building, such as: better-performing building materials, more fire-resistant components and assemblies, and safer and more reliable electrical systems.

Further, one does not find small residential buildings, or for that matter larger ones, receiving the special attention of architects with respect to fire safety design. In general, there is little interest in providing more fire exits, shortening travel distances to exits, reducing the size of compartments, or in the creation of areas of refuge, beyond what is demanded in prevailing building codes.

Rather, we are more likely to find the building designer primarily interested in attaining a high efficiency of space utilization, and in style, amenities, and aesthetics that will sell the building to owners and tenants. A common response of architects to this accusation is that fire protection is the responsibility of others. But the leadership position that the architect holds in the design process, and the special relationship that the architect has with building owners, actually place them in a most influential position with regard to levels of fire safety that can be achieved.

Figure 5.1. Fire tests conducted in "unengineered" row frame residential buildings in New York City—Fires are fast and deadly.

When considering fires in buildings, it is of some interest that incendiary and suspicious fires account for the largest share of direct property damage, and for perhaps as many as 1000 deaths annually. While there is some argument that "set" fires and the actions of terrorists are beyond the scope of ordinary building design, fire statistics seem to belie that position, suggesting instead that contemporary public buildings be, as much as possible, hardened against such assaults. It should also be of interest to designers that almost half of all deaths due to incendiary incidents result from fires set in the means of egress.

Figure 5.2. Effects of a fast-moving exposure fires in "unengineered" residential buildings—sixteen buildings involved in a single fire incident.

5.4.3 Fire Fighter Deaths and Injuries

Discussion of fire deaths and injuries would not be complete without noting that each year over 100 fire fighters lose their lives in fires, and line-of-duty injuries number in the high thousands. In recent years in New York City, which has a fire fighting force of approximately 12,000, lost time resulting from on-duty injuries exceeds 100,000 man-days annually.

Recent reports of fire fighter deaths by type of duty show that approximately 53 percent result from firefighting operations, 23 percent are attributed to response and return runs, and the remaining percentage occurs in the performance of other than firefighting duties.

Deaths and injuries to fire fighters, in terms of their numbers, the circumstances under which they occur, or building designs and operations that may be directly causative or contributory, are rarely considered by the design professions. For that matter, neither building departments nor the insurance community appear to have much interest in matters that affect fire fighter safety.

Thus, we find that, both in the architectural design and engineering of buildings, envelopes that render ventilation of fires extremely difficult, and which make effective and safe laddering and entry through intricate facade grillages impossible. We also find that stair and elevator configurations and locations are hardly ever designed with recognition of the need for fire fighters to gain quick and easy access to the seat of fire.

While the evacuation of occupants from buildings normally takes place within the first few minutes of a fire, fire fighters remain within the fire building conducting search and rescue operations and attempting to control the fire for much longer periods, and well into the time frame when structural components, partitions, ceilings, doors, and windows may fail.

They also operate on and close to the building as they mount exterior attacks, man rescue ladders, and ventilate roofs. Because building codes do not generally acknowledge fire fighter needs, designers should be especially alert to those details and elements of the structure, both interior and exterior, which are critical to fire fighters and which may be expected to fail during the first two hours or so of exposure to fire temperatures and pressures.

One other concern, with regard to excessive structure loading that might be experienced during a fire incident, is that of water loading. Roof and floor slabs have failed where drainage facilities cannot carry off flows from hose lines. Such accumulations can result in over 5 pounds per square foot per inch of impounded water being added to deck loads. Needless to say, there is no reason why such water loading on roofs cannot occur during periods when there is already a significant snow and/ or ice load acting and normal drainage is impeded.

5.4.4 Property Damage

Structural fires account for approximately 88 percent of all property losses due to fires. In 1984, such losses amounted to approximately $5.8 billion of the total $6.7 billion resulting from all fires.

Although it is argued that measures that reduce property losses also reduce casualties, and vice-versa, this perspective is far too simplistic to properly address the problems in either domain. In fact, there are circumstances under which measures directed to the preservation of the structure may prove to be counterproductive, with respect to occupant safety and risks to fire fighters.

For example, a labyrinth of small compartments, which may protect the structure, may make occupant escape and fire fighter access difficult. Also, highly articulated smoke-exhaust systems, if not properly designed and utilized, while effectively carrying smoke and heat out of a building (and reducing smoke and fire damage to the structure), may expose occupants to deadly smoke and gases, feed the fire with air, and cause it to move in the direction of fire fighters.

5.5 SUMMARY

This chapter has established the broad contexts in which issues of fire safety of building occupants are set. Challenges presented by tragic fires that have occurred during this century in various types of occupancies have been identified, and some of the perceptions and attitudes of building design professionals have been touched upon. A number of general fire statistics have been reviewed, and some of the costs of unwanted fire, in terms of civilian and fire fighter casualties suffered and levels of property damage incurred, have been presented.

The chapter that follows will discuss the fundamental relationships which describe the behavior and performance of materials and elements of building construction under fire conditions. Chapter 7 presents some of the current issues facing fire protection designers, and considers the available measures for enhancing safety from fire in buildings.

REFERENCES

Backes, Nancy. *Great Fires of America.* Waukesha, WI: Country Beautiful Corp. 1973.

Ross, Steven S. *Construction Disasters.* New York: McGraw Hill Book Co. 1984.

Bell, James R. Investigation report of fire at Stouffer's Inn of Westchester. *Fire Journal,* NFPA. May 1982.

Best, R. L. Reconstruction of a tragedy—The Beverly Hills Supper Club fire. *Fire Journal*, NFPA. Jan. 1978.

Breen, David E. The MGM fire and the spread of flames. *SFPE Bulletin*. January 1981.

Emmons, Howard W. Analysis of tragedy. *NFPA Fire Technology*. May 1983.

National Bureau of Standards. *An Engineering Analysis of the Early Stages of Fire Development—The Fire at the Dupont Plaza Hotel and Casino*, December 31, 1986. Washington, D.C.: Harold E. Nelson National Bureau of Standards, Center for Fire Research. NBSIR 87-3560. April 1987.

National Fire Protection Association. The Fire Almanac. Quincy MA: NFPA. 1984.

Chapter 6
Fire Hazard: Fire Phenomena and Effects

Paul R. De Cicco, P.E.
Polytechnic University

6.1 DEFINITION OF FIRE

Although the previous discussion of large historical fires and recent fire statistics clearly depicts their more destructive manifestations, there are instances when a more precise definition of fire becomes essential. This occurs in the scientific analysis of ignition phenomena and in connection with litigations, in which it is necessary to establish specific causes of casualties and property damage. Insurance companies will often challenge claims for losses on the basis of whether a fire actually occurred or not.

Webster's Unabridged Twentieth Century Dictionary gives the definition: "The active principle of burning, characterized by the heat and light of combustion."

The American Heritage Dictionary of the English Language offers: "A rapid, persistent chemical reaction that releases light and heat; especially, the exothermic combination of a combustible substance with oxygen."

A more technical definition is given in terms of combustion in the *Fire Protection Handbook* (1986) published by the National Fire Prevention Association. It states:

> Combustion is an exothermic, self-sustaining reaction involving a solid, liquid, and/or gas-phase fuel. The process is usually (but not necessarily) associated with the oxidation of a fuel by atmospheric oxygen with the emission of light. Generally, solid and liquid fuels vaporize before burning. Sometimes a solid can burn directly by glowing combustion or smoldering. Gas-phase combustion usually occurs with a visible flame. If the process is confined so that a rapid pressure rise occurs, it is called an explosion.

Electrical malfunction also often comes into contention with regard to whether a fire has occurred or not. In one case, a court stated:

the burning or charring of a wire carrying electric current or occurring during or accompanying an overload of current must be regarded as an electrical injury especially when, as here, there is an absence of evidence showing such burning or charring to be fire as distinguished, if that can be, from electrical injury.

In this case, demonstration that there was heat damage was not sufficient to indicate that a fire had taken place.

Since ordinary structural materials are rarely the cause of ignition or even the first material ignited (although exposed wood may affect fire spread and fire extension characteristics), the structural engineer is little concerned with fine definitions of what fire is. In fact, by the time structural components become involved in a building fire, there is little question that a fire has occurred.

It should, however, be noted in this same context that the long-time exposure to heat, of structural members composed of combustible materials or faced with combustible materials, will often result in their ignition temperature being considerably lowered. Under such circumstances, these items might indeed be the first materials ignited. Exposure of combustible structural members to heating units, hot water pipes, and overloaded (or poorly ventilated) electrical wiring, have all been associated with costly building fires.

6.2   THE FIRE TRIANGLE

The three well-known requirements for fire, referred to as the fire triangle, are: a source of fuel, a supply of oxygen, and an ignition source. Each of these items can readily be found in many forms, quantities, and degrees of hazard in all buildings. Most fire prevention measures are implemented through actions and mechanisms that remove one or more of these necessary ingredients.

6.2.1 Fuel

Fuel sources are present and readily available for ignition and combustion in finishes and decorations used on walls, floors, and ceilings, and in furnishings and other combustible contents brought into buildings. These are usually the first materials ignited in unwanted fires. Other materials that may be the first to ignite are liquid and gaseous fuels for cooking, space heating, and for power generation. While it is relatively uncommon for structural components to be among first ignited materials, electrical malfunctions, such as overheating of wires, short circuits, and arcing phenomena are sometimes responsible for direct ignition of wood and plastic elements of structures.

The readiness with which fuels will ignite, and the nature and quantities of their products of combustion (including heat, smoke, and toxic gases) are all critical to the outcome of fires in buildings and to the safety of occupants. Other fuel combustion characteristics that are of great importance to life safety are the rapidity with which flames will spread across fuel surfaces and the rate at which heat will be released.

6.2.2 Oxygen

Oxygen is usually made available through the direct supply of air to a fire. Air may be available in only limited quantities, such as in the case of small, enclosed spaces without window or door openings; in deep materials, such as upholstered furniture or within piles of paper, wood, or textile materials; or in enclosed vessels and containers of various types. More substantial ventilation through windows, doors, and other openings, or supplied by mechanical systems, will greatly increase fire growth and extension.

6.2.3 Sources of Ignition

Ignition sources in buildings abound. They exist as items used by occupants, such as smoking materials and electrical appliances of all kinds; as tools, including torches, electric drills, and other power devices; as elements of building service systems, such as electrical wiring, motors, transformers, and a wide variety of machinery used to provide vertical transportation and to heat and cool the building. Ignition sources may also be brought into buildings as incendiary devices. The frequency of arson fires in recent years, reported to be on the order of 20 percent of all fires in the United States, can no longer be totally ignored by designers responsible for occupant safety from fire.

It has been estimated that 1570 American die each year as a result of cigarette-ignited fires, and that over 7000 are seriously injured. Property damage from this source of ignition has been given as $390,000,000. It is also of some interest to note that new minimum standards for fire cigarette safety are being proposed in congress, and that prototype cigarettes based on reducing circumference, decreasing tobacco density, using less porous paper, and reducing citrate additives have been developed under experimental programs.

While most of these ignition circumstances lie outside of the influence of building designers, they greatly affect where fires are likely to occur, their rates of growth, and how they may expose structural elements of a building.

6.3 FIRE PROPERTIES OF MATERIALS

As mentioned above, there are a number of properties of materials of construction and building contents that establish how they will behave and perform during a fire, and determine their inherent tendencies to become involved in fire in the first place. These properties include ignitability, flame spread, smoke production rate, energy contained, rate of energy release under fire conditions, and other physical and chemical attributes that dictate the roles that these materials are likely to play during a fire incident.

Some of the important properties of a number of common materials of construction are discussed below.

6.3.1 Wood

Wood is the most commonly used construction material for framing, sheathing, flooring, roofing, and siding in smaller residential buildings, and is also used in large quantities in smaller commercial and mercantile structures. In larger structures, it is frequently used in flooring, wall panelling, and for a wide variety of structural and decorative purposes. It has been reported that the per capita use of forest products in the United States is approximately twice that of all metals on a per-weight basis.

Wood generally ignites readily, contains relatively high heats of combustion and emits dangerous amounts of smoke including carbon monoxide and other noxious gases under ordinary fire conditions. Since it is usually protected by paints or other types of combustible coatings, wood must be considered one of the serious fire hazards in any building and should be used accordingly.

As a structural material, wood is used in a wide range of elements including beams, stringers, posts, decking, purlins, glue laminated timber, plywood sheathing, poles and piling. It is also used in wood trusses found in churches, residences and a wide variety of commercial buildings.

Tables 6.1 and 6.2, which follow, give ignition temperatures for representative woods and the heats of combustion for a number of materials.

Wood exposed to continuous high temperatures may ignite at temperatures considerably lower than those listed. Some researchers have reported that long-time temperature exposures exceeding 212 degrees Fahrenheit are likely to be reflected in such low-temperature ignitions.

The event of "flashover" in a compartment fire, which occurs at that instant when heated, uncombusted gases in the upper part of the space, as well as other solid fuels in the room, reach ignition temperatures and cause the room to literally flash into flame, is extremely critical in fire episodes. A fire that may already be growing rapidly, suddenly fills the space at flashover, leaving little chance for the escape of occupants. The course of the fire from that point on, and the resulting intensity of heat exposure of room surfaces and structural components, can critically tax the integrity of the structure.

Work performed at the National Bureau of Standards indicates that if the heat release in a fire (in megawatts) exceeds $0.75\ A\sqrt{h}$, where A and h are the areas and heights of windows and door openings, respectively, the room is likely to flashover. Wood prod-

Table 6.1 Ignition temperatures of wood (NBS).

WOOD	SELF-IGNITION TEMPERATURES	
	°F	C
Short leaf pine	442	228
Long leaf pine	446	230
Douglas fir	500	260
Spruce	502	261
White pine	507	264

Table 6.2 Heats of combustion of wood and other materials (all woods on dry basis) (NFPA, 1986).

SUBSTANCE	HEATING VALUE IN BTU PER LB
Wood sawdust (oak)	8,493
Wood sawdust (pine)	9,676
Wood shavings	8,248
Wood bark (fir)	9,496
Corrugated fiber carton	5,970
Newspaper	7,883
Wrapping paper	7,106
Petroleum coke	15,800
Asphalt	17,158
Oil (cottonseed)	17,100
Oil (paraffin)	17,640

ucts having moisture contents of 8 percent were observed to have heat release rates between 90 and 170 kw/m^2.

As noted earlier, the amount of air (oxygen) supplied to a fire is essential to the completeness of combustion that will occur. The vigor of flaming, combustion, and the amount of carbon monoxide produced, are therefore dependent on ventilation conditions, the size of the fire compartment, and the chemical composition of materials that are burning.

The volume of air required for complete combustion of solid fuels may be estimated as:

$$V = 147 \ [C + 3(H - O/8)]$$

where V equals the volume of air in cubic feet, and C, H, and O, are the parts by weight of carbon, hydrogen, and oxygen in one pound of fuel, respectively.

The treatment of wood, plastics, and fiber materials to reduce flammability is commonly required for sheathing and paneling, carpeting, wall coverings, and other textile products used in buildings. When properly performed, such treatment can retard or delay ignition of these materials. It is, however, important for designers to not only specify such fire retardance, but also to have some reasonable knowledge as to how such materials will behave under fire conditions.

Designers should also expect that when such "protected" and "made safe" materials are exposed to sufficiently high temperatures for extended periods of time, they will burn and contribute very much the same amounts of heat energy and perhaps even greater amounts of smoke and toxic gases as they would without treatment. The point here is that if the fire is large enough, structural components will be spared little in the way of temperature stresses, and occupants will be highly exposed if there are significant quantities of such combustibles, even if they are treated.

Often, architects and engineers, in selecting and placing materials, fail to appreciate that the quantity, configuration, and surface area of a potential fuel are equally as important as the type of material. For example, a heavy oak conference table in the center of a room may represent more fuel in terms of its mass, but a lesser mass of similar material used to cover all of the walls of the room will present a far greater fire hazard.

This analogy also holds true for the behavior of many solid plastics, when compared with their expanded foam counterparts. That is, considerably less mass used as thermal insulation can make a relatively large contribution to the rate of energy release and to the growth of fire. For this reason, the role of plastics in roof insulation and covering and in exterior wall insulation merits special attention, when considering external spread of fire.

As has been mentioned earlier, the rate at which materials release heat energy is an extremely important attribute of fuels that become involved in building fires. The safe escape of occupants, assuming that there is adequate and prompt warning, depends much on how fast a fire grows, and on how soon conditions become untenable along the path of egress.

For example, after a slightly slower release during the first minute or less, the rate of heat release from burning polystyrene has been reported to peak at three times the rate of a similar volume of wood. The rapid release of heat drives fires to earlier flashover times, which not only heightens the danger to occupants attempting to escape or to hold fast in a "safe" area, but also deprives fire fighters of the advantage of addressing a more moderate fire condition on their arrival at the scene.

6.3.2 Concrete

Next to wood, concrete is perhaps the most commonly used construction material in the United States. Its strength and other performance characteristics and qualities are constantly being improved, and because it is also used to protect structural steel, it is assured of a continuing important place in building construction.

The strength and stiffness properties of concrete decrease at temperatures above approximately 200 degrees Fahrenheit. Exposure to temperatures above 900 degrees may result in spalling, and quartz (a major constituent) changes state and expands significantly at temperatures above approximately 1000 degrees Fahrenheit, greatly affecting its structure.

Differential expansion of concrete materials as a part of its mass is exposed to high fire temperatures, while other parts and surfaces remain cool, is also responsible for spalling and cracking. This, in turn, exposes reinforcing steel to high temperatures, with a rapid and appreciable loss of strength.

G.W. Washa, in the *Concrete Construction Handbook* (1974), states that the endurance of concrete walls depends primarily upon their thickness, with a doubling of fire resistance observed for an increase of 35–40 percent in wall thickness. In addition, the type of construction (solid concrete wall versus a hollow, masonry unit wall; full bedding of masonry units versus bedding of the face shells only; and thickness of reinforcing covering), the type of aggregate (lightweight aggregates, including inflated clays, shales, and slags are superior to natural aggregates), and quality of concrete (greater cement content), are also important factors concerning performance under fire exposure.

It is important to note that concrete, after exposure to fire temperatures, which may exceed 2000 degrees Fahrenheit, exhibits a decrease in strength of up to 35 percent for hollow test units, made with the best aggregates. Washa also reports moduli of elasticity prior to exposure, from 200,000 to 750,000 psi, and after exposure values of from 100,000 to 300,000 psi. This suggests the need for careful inspection and testing of buildings in which serious fires have occurred.

5.3.3 Steel

The tensile strength of ordinary structural grade steel begins to decrease as temperatures increase above approximately 500 degrees Fahrenheit. At temperatures of 1100 degrees, ordinary structural steel may have less than 50 percent of its original strength. During a fire, the manner of heating and maximum temperatures reached in structural steel members may be extremely nonuniform in nature, with differences of over 600 degrees Fahrenheit observed between upper and lower flanges under standard test conditions.

For this reason, deformation may become of greater importance than temperature, and performance can be unpredictable. During operations to control building fires, structural steel may also be subjected to rapid cooling by fire hoses. This adds another degree of uncertainty regarding its behavior during fire incidents (American Iron and Steel Institute, 1971).

The general comments made above with regard to the properties of steel are somewhat simplistic. In fact, the specific form, mass, and conditions of exposure of steel structural elements and assemblies have much to do with their performance. For example, solid web steel beams or girders and open web steel joists will behave differently when exposed to fire, and the grade of steel, whether it has been cold worked, and differences between test conditions and actual fire conditions are also significant.

In addition, the shape of fire protection applied, as well as the magnitude of imposed loads on steel members and constructions, are important factors in the prediction of the performance of steel elements at elevated temperatures.

Restraint conditions of structural members and assemblies in buildings under fire conditions are also difficult to predict, and may be quite unlike those used in laboratory tests. Magnitudes and distributions of stresses caused by elevated temperatures on members under restraint have significant effects on the behavior of structural assemblies. Restraint conditions in fire tests are usually specified for floor and roof assemblies, and for concrete and steel beams. Both restrained and unrestrained ratings may be given with conditions of acceptance, including load failure, heat transmission on unexposed surface or ignition of cotton waste, and temperature criteria.

The American Iron and Steel Institute, in its publication *Designing Fire Protection For Steel Beams* (1971) discusses the subjects of restraint, continuity, and redundancy, and makes reference to an appendix to the ASTM E119 Standard Fire Test, entitled "Guide for Determining Conditions of Restraint for Floor and Roof Assemblies."

This appendix defines restrained assemblies as follows:

Floor and roof assemblies and individual beams in buildings shall be considered restrained when the surrounding or supporting structure is capable of resisting substantial thermal expansion throughout the range of anticipated elevated temperature. Construction not complying with this definition is assumed to be free to rotate and expand and shall therefore be considered unrestrained.

Figure 6.1a shows the standard time-temperature curve used in the ASTM E-119 Fire Test, and suggests the conditions that structural steel must be able to withstand. Figure 6.1b shows how the

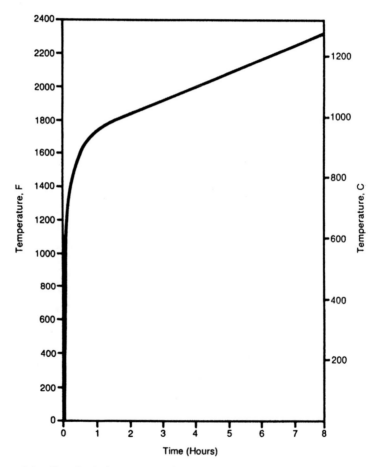

Figure 6.1a. Standard time-temperature curve used in ASTM E 119 Fire Test Reprinted from *Fire Protection Through Modern Building Codes*, 5th ed. with permission. © 1981 AISI).

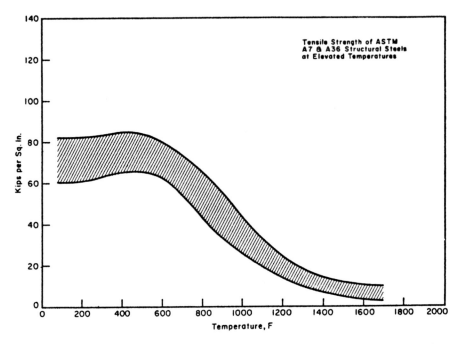

Figure 6.1b. Tensile strength versus temperature A7 and A36 structural steels (Reprinted from *Fire Protection Through Modern Building Codes,* 5th ed. with permission. © 1981 AISI).

tensile strengths of A7 and A36 structural steels vary with temperature. Figure 6.1c shows how the *yield strengths* of A7 and A36 structural steel vary with temperature.

Fire Protection of Steel. Because of the loss of strength that results from the exposure of steel to fire temperatures, structural members and assemblies must be protected. Materials used for this purpose must be able to resist the transfer of heat, must maintain stability under fire exposure conditions, and must be durable under environmental and other conditions of use.

The thickness and density of coatings must be controlled, and adherence assured for the life of the building. Fire experience well documents instances where fireproofing materials failed to adhere to corroded steel surfaces, with disastrous results during severe fires.

Commonly used materials for the protection of steel include gypsum, which may be used alone or with aggregate materials, such as

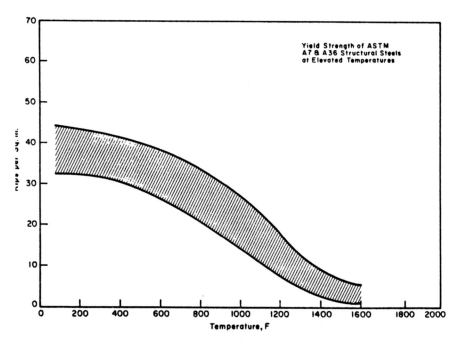

Figure 6.1c. Yield strength versus temperature A7 and A36 structural steels (Reprinted from *Fire Protection Through Modern Building Codes*, 5th ed. with permission. © 1981 AISI).

lightweight insulating vermiculite or perlite. Spray-on materials consisting of mineral fibers or cementitious materials are also frequently used. As an alternative to coatings, noncombustible ceiling systems may be employed. Here it is critical that ceiling spaces do not contain combustible materials.

Concrete containing limestone, trap rock, slag, burnt clay, and lightweight aggregates is commonly used as fire protection on steel. Poured concrete protection must be reinforced with wire mesh. To a lesser degree, masonry, intumescent coatings, and sheet steel membranes are also in use to protect structural steel.

Masonry materials including brick, concrete blocks, clay, and tile offer good fire-resistive properties, and are often used to protect steel. Material thickness and the quality of workmanship are the two most important factors affecting their performance. When they are used for the protection of steel, they must be firmly secured and bonded at horizontal joints using clips, outside tie wires, or other means.

Field testing procedures given in ASTM standards, and specifications offered by manufacturers and other agencies and organizations interested in the protection of steel against fire, provide guidance concerning materials and their application. In addition, controlled inspections by an architect or engineer are required by most building codes to assure the quality of such measures.

Exposed Steel.   More recently, and in greater numbers of building projects, steel is being left exposed without the benefit of fire protection. In these cases, it must be rigorously shown that, because of the location of the steel, with regard to the proximity of fuel loads and anticipated fire dynamics, temperatures will be considerably below that which might cause the failure of structural members and assemblies. Since exposed steel is often associated with major members and assemblies that support large roof structures, it is imperative that possible fire scenarios be carefully considered, as failure under fire conditions will be extremely serious. Such exposed steel may also be protected by sprinkler systems, and some codes require that this be done.

Exposed steel is being used successfully in many types of structures. It is increasingly found in garages; as columns, spandrels, and diagonal members on the exteriors of high-rise buildings; and in a host of other applications. It is also important to note that increasingly reliable analytical methods and fire testing procedures have become available for establishing the safe use of such construction.

Underwriters' Laboratories, the national building code organizations, and the technical arms of the steel industry provide design information and details for the protection of structural assemblies.

Steel Columns.   Columns, like other major structural elements, while critical to the support of floors and roofs and to the overall stability of the building, usually do not fail under fire exposure during the period when the evacuation of occupants will be taking place. On the other hand, columns that have deteriorated over long use, through corrosion, and columns which have suffered mechanical damage from impact, or have been weakened by inappropriate add-on constructions, are much more susceptible to failures under fire stress. The unexpected collapse of floors, roofs, and walls are the nemesis of fire fighters, and generally stem from these kinds of conditions.

In many instances of early and premature building collapse during fires, it is found that older buildings, still capable of sustaining normal loading, cannot endure the additional stresses imposed by thermal deflections and deformations, or (in the case of wood and concrete members) the loss of material resulting from burning or spalling.

In the case of columns, it must be remembered that connections at the tops will be exposed to the highest fire temperatures, often exceeding 2000 degrees Fahrenheit, which normally occur at ceilings and within concealed spaces.

Steel Beams, Girders, and Joists.   The protection of steel beams and girders is usually accomplished by encasement, by suspended ceilings, or both. Steel joists are protected by suspended ceilings. With respect to the life safety of occupants, it is necessary to observe the usual cautions concerning the need for good workmanship in spaces that will later be concealed, and the importance of maintaining such concealed spaces free of combustible materials. It should also be recognized that, while not principal load-bearing constructions, suspended ceilings and the spaces that they enclose are extremely vulnerable to early fire effects. Since they may be exposed to fire from either or both sides, and their premature failure can represent a hazard to occupants.

Ceilings are often added or replaced after the original construction, and therefore materials and designs vary widely. In addition, aesthetic and acoustical attributes are likely to receive priority over strength and behavior under fire conditions. From a fire safety point of view, suspended ceilings that create concealed spaces where combustible materials may be stored, where penetrations of rated walls may go unseen, and which provide passages through which fire may travel across the tops of occupied spaces, should be considered with great care in fire-safe design.

5.3.4 Walls and Partitions

The fire resistance of walls and partitions must be such as to prevent the passage of flames, resist temperature increases that would ignite combustible materials on the unexposed surface of the assembly, and limit the temperature rise on the unexposed side. Ordinarily, the wall must also be able to resist the impact and cooling effects of a hose stream played on its surface after prescribed exposure to a test fire. Load-bearing walls must be able to pass these tests while under full design load.

Historically, the safety of occupants has seldom been jeopardized by the failure of walls per se. This is because the critical escape time of occupants is generally much less than the time necessary for fire to penetrate most wall constructions. Wall failures leading to casualties are more likely to involve the passage of smoke and fire through openings and penetrations of interior walls, in connection with constructions in which walls are not properly carried through concealed communicating spaces in ceilings, or are not joined with exterior walls in such a way as to preclude fire passing around them from one compartment to the next. Ineffective fire and smoke dampers where ducts pass through interior walls, open doors between compartments during fires, the omission or improper installation of fire-stopping material in the walls of shafts and pipe and electrical chases, and the use of combustible finishes are more likely to result in harm to occupants than is the direct passage of fire through rated surfaces.

Both the architect and design engineer must exercise extreme care with respect to inspection of work in progress to be certain that wall construction is carried out to meet fire requirements, especially with regard to details that may affect the passage of fire or smoke and which, on occupancy, may be concealed from view.

6.3.5 Plastics

The ASTM definition of "plastic" is:

> A material that contains as an essential ingredient one or more organic polymeric substances of large molecular weight, is solid in its finished state, and at some stage in its manufacture or processing into finished articles can be shaped by flow.

Polymers are divided into three classes, which include: elastomers, thermosets, and thermoplastics. In addition to the polymer, finished plastics may contain plasticizers, colorants, fillers, stabiizers, and lubricants. An increasing use of reinforced plastics for structural applications has focussed increasing attention to reinforcing materials and their performance under fire conditions. There is probably no other domain, regarding the testing of construction materials and assemblies, where the caveat that these items should be tested in environments and under physical conditions matching their actual use is more critical.

The NFPA *Fire Protection Handbook* lists approximately 30 major groups of plastics, and gives brief descriptions of their uses and general behavior attributes.

The use of plastics in construction materials and assemblies for walls, roofs, flooring, and plumbing, as well as for a wide variety of finishes, furnishings, and other appliances, accessories, and packaging materials which find their way into buildings, has become a major concern to fire protection engineers over the past four decades.

The diversity of the types and formulations of plastics in use is broad, with these materials displaying an extremely wide range of properties and attributes relating to ignitability, flame spread, heat release, smoke production, and toxicity. How they are manufactured and conditioned, the type and effectiveness of fire-retardant treatment they undergo, and whether they are used within or on the exterior of a building also have important impacts on their performance under fire conditions. In addition, the fire performance of coatings and membranes used to protect these materials, and thermal barriers used to separate them from interior spaces, are critical to their safe use in buildings. Finally, the quantities used and their installed configurations can have significant effects on their behavior at fire temperatures.

Test methods and standards for plastics are issued by NFPA, the American Society for Testing and Materials (ASTM), Underwriters' Laboratories, and a number of federal agencies, including the Department of Transportation (DOT), the U.S. Consumer Product Safety Commission, and others.

Current building codes are generally cautious and conservative regarding the uses of plastics as construction materials, but there is much less control with regard to their use in decorations, furnishings, and other contents that figure heavily in fires and the resulting deaths and injuries.

In general, and unless proven safe through reliable test documentation and actual fire experience, the use of plastics in buildings should be regarded with great care.

6.4 FIRE SAFETY IN BUILDINGS, BY DESIGN

Much of the current wisdom concerning fire safety in buildings is reflected in two instruments that are generally required by building and fire codes in the United States. These are the Building Fire Safety Plan and the Building Fire Protection Plan. The former addresses what is expected of occupants and building staff charged with fire safety, and the latter describes devices, equipment, and the modes of operation of engineered fire protection systems incorporated in the building.

6.4.1 Fire Safety Plan

Over the past two decades or so, most municipalities have required, as part of fire safety regulations, that every public-use building, and particularly high-rise structures over 75 or 100 feet in height (considered to be above reachable height of fire ladders), have a Fire Safety Plan. Such plans specify that there be a Fire Safety Director, fire wardens, and predetermined procedures for alerting and assisting occupants in evacuating during fire emergencies. Fire drills of various types, conducted at regular intervals, are also mandated.

While such plans will not overcome serious deficiencies in building design, it must be recognized that even the most complete and advanced measures for fire protection in buildings are apt to be of no avail if confusion, panic, misdirection, and interference with rational escape procedures occur.

It is of importance for the designers of all contributing disciplines to be familiar with activities that occur during fire emergencies. This includes: what occupants will be directed to do, paths of escape they will take, typical actions of both building fire brigades and regular fire services, and procedures to be followed by maintenance staff with regard to the control of air handling systems, the opening or closing of doors, and the operation of gravity ventilation facilities.

The lack of an adequate Fire Safety Plan, or failure to implement such a plan, were contributing factors in many deaths and injuries in the previously cited fires. The manner and means of supervising fire protection systems, and insuring that occupants are either able to evacuate buildings safely or can remain in safe areas until they are rescued or the emergency is over, are generally not uppermost in the priorities of structural designers. But these professionals should be aware of realities long known in the fire community, which are that the behavior of building staff and occupants during fire emergencies may be far from what is rational and desired, and that fire service response time (the time when water for extinguishment in sufficient quantities can be brought to bear to a fire on an upper floor of a high-rise building) may be 30 minutes or more after notification, even in the best protected communities.

Under these circumstances, the structural integrity of the building and the fire resistance of elements considered to be mere details may be called on to perform under extreme thermal and mechanical conditions. For example, the quality and fitting of doors, the performance of fire barriers and draft curtains, and the adequacy of fire-stopping materials, are all critical during building fires.

6.4.2 Fire Protection Plan

When plans for building projects of any size are submitted for examination, it is usually required that the architect, with the assistance of a fire protection engineer, file a detailed statement describing all fire-protection features to be included in the structure. Any deviations from required provisions of governing fire and building codes must be noted, with an explanation of how such deviations are to be satisfied by alternate measures. In a recent submission in New York City for an unusually large, multi-use building, approximately 1000 feet in height and containing over three million square feet of space, the "Fire Safety Plan" included:

1. The names of all architects and engineers, including fire protection engineers, special consultants, and construction management professionals engaged in the project.
2. A description of the site, including all bounding streets and information concerning building spaces above and below ground.
3. The street addresses of different parts of the project and the locations of all entry points.

4. Details of below-ground spaces, including the capacity and number of levels occupied by parking garages, and how these facilities are separated from other occupied spaces.

5. The location of all lobbies, loading facilities, and fire service access points.

6. The locations of all major building occupancies and descriptions of each use, including public assembly spaces such as theatres, restaurants, auditoria, and atria.

7. A detailed schedule of all elevators, showing the type (passenger or service), capacity, floors served, special controls such as fireman service, provisions for recall during fire emergencies, and in-car emergency communications; standby and emergency power provisions and status and control panels; a schedule of escalators, dumbwaiters, trash chutes, and the means for their protection during fires.

8. Descriptions of all major fire protection measures including: sprinkler systems, pumps, standpipes, a listing of sprinkler densities in various areas, and identification of spaces that will not be sprinklered, including reasons and alternative measures to be used; a schedule of all fire stairs, indicating floors served and whether they are smoke protected; descriptions of early-warning systems in each major area of the building and an account of how they will function during a fire; a description of emergency power provisions for light and other critical devices and facilities during a fire emergency; the types and locations of portable fire extinguishers; the types, locations, and manner of operation of halon systems; door unlocking systems and descriptions of automatic operating modes under various detector, alarm, and sprinkler flow conditions.

9. The location or locations of fire control consoles, including the types of annunciators, communication options, and controls that will be available, including itemized lists of all damper and air handling functions that may be controlled during a fire, and the locations of remote annunciators in security and building management areas.

10. A description of the building organization that will manage fire safety, including the Fire Safety Director, deputies, floor wardens, and fire brigade, in accordance with requirements specified in the building code.

11. A description of the basic construction materials, including foundations, superstructure framing, wall systems, windows, and glazing.
12. Descriptions of how each major component and system is to function during fire emergency.

A special caution for building designers concerned with the safety of occupants is that it is generally bad practice to hastily omit, alter, or attempt to "trade-off" code provisions. Code language generally does not provide an explanation of the reasons behind requirements. Often, what appear to be obvious and easy to replace code provisions are, in reality, more complex in their intent and purpose. Omission, or substitution of hastily conceived "fixes" in their place, can result in severe consequences during the time of a fire.

For example, sprinklers are sometimes sacrificed (with, perhaps, an additional smoke detector substituted) in locations where it might be inconvenient to bring in water lines. Fire-stopping in utility chases is often postponed because "more work may need to be done later." The "unbundling" of combinations of mutually supportive fire-protection measures, such as concurrent requirements for compartmentation and protection of fire stairs from smoke, which together provide significant safety, while either one standing alone contributes little of the protection intended, are further examples of ill-advised transgressions of code provisions.

Similarly, complete sprinklerization, while of great and undisputed value, does not remove all life safety hazards, since fires may produce dangerous levels of smoke before sprinklers operate. In addition, under certain fire conditions, as in the case of smoldering fires or fires sheltered from sprinkler heads, sprinkler flow may be delayed or not occur at all.

It is also known that the concentration of carbon monoxide, the most common and most significant of all toxic products of combustion, may increase after sprinklers actuate and thus interfere with complete combustion. The most frequent cause for sprinkler systems failure is the lack of a reliable water supply. This usually results from the improper closing of valves, and there is no absolute protection against this occurrence.

It is also relevant to observe that, recently, there has been a tendency to remove or reduce other safety measures in sprinklered buildings. While this practice may be quite tempting in the interests of economy, they should be cautiously approached, even if codes per se (or the discretion of regulatory authorities) permit such actions.

6.4.3 The Building Design Team

The building industry is extremely large, complex, and fragmented, and almost all of its many components influence the quality of fire safety of buildings in one way or another. Ultimately, how a building will perform under fire conditions is much determined by the owners, who provide the funding, and developers, who assemble the building site, engage necessary design and construction forces, and interface with regulatory agencies and prospective tenants.

The concept and value of the "design team" has long been recognized in the construction industry. To some extent, its perception as a deliberative body, made up of major contributors to the design and engineering of a building, discussing and establishing priorities and coordinating the work of the various technological disciplines involved, it perhaps more of a promise than a reality, in terms of how it actually functions with regard to life-safety issues. The concept is, however, important enough to warrant consideration here.

The building design team nominally is made up of the principal contributors to the project. As such, the architect usually acts as coordinator, unless some other overall design manager has been assigned this responsibility. The structural, mechanical, and electrical engineers responsible for the various building systems, and the fire protection engineer (who may be part of the mechanical engineering organization) make up the engineering contingent of the team. In recent years (on larger projects), there will also be found a construction manager, whose responsibility is to keep the project on time and within budget. It is sometimes alleged that construction managers sacrifice both quality and safety in the interest of discharging these responsibilities.

The perception that the design team, represented always by the same individuals, is a constant monitoring force is only partly true. In fact, there are many substitutions of people assigned to attend conferences and, depending upon the subject matter under consideration, the faces around the meeting table may change often. This lack of continuity can result in important issues "falling between the cracks," as the transfer of information from one representative to another in a specific area of design may be imperfect. This has been observed by the author in connection with the planning and design of fire detection and alarm systems in a number of large and complex buildings.

One other aspect of design team operation that needs to be recognized and addressed in the interest of the safety of building occupants from fire has to do with the chain of command. Although it may be argued that each technical representative — architect, structural engineer, mechanical engineer, electrical engineer, etc. — is a master in his own right, when fire safety recommendations proposed are in addition to code requirements, they are often regarded as "suggestions" and deemed to be luxuries. As such, they are often denied by architects, owners, and project construction managers. In effect, they are traded for more usable space or other amenities that make the project more saleable.

Some examples of fire safety enhancements often rejected because they exceed code requirements are: sprinklering of buildings, sprinklering of duplex apartments in residential buildings, two-way communication systems for occupied spaces of commercial and residential occupancies, increased water supply and sprinkler densities in areas of uncertain hazard, sprinklering of atrium floors, providing additional or oversized exits in large assembly places, adding strategically located fire fighter access points to large interior spaces, direct exits to the exterior, and smoke-proof stair towers.

6.5 BUILDING CODES

Somewhat of a paradox exists in the United States with regard to unwanted fire. For, with construction technologies second to none, and building codes considered to be the most stringent found anywhere in the world, the U.S. continues to have the worst fire statistics, with regard to both casualties and property damage suffered annually.

A recent report on worldwide fire services indicates a wide range of figures for the numbers of fire personnel, reported fires, and fire deaths in major cities (populations one-half to three million). Cities in the United States have from one and one-half to two times as many fire-fighting personnel per capita as do their overseas counter-parts. At the same time, foreign cities have from 1.3 to 1.6 times the personnel per fire. The answer to the obvious question as to whether we in the United States have too few fire fighters or too many fires lies in the statistic that the U.S. has three times the number of fires per capita than most of the rest of the world's industrialized nations.

Two reasons are offered for this situation. The first is that the United States is by far the heaviest consumer of electrical energy, and that many fires stem directly or indirectly from this fact. Specifi-cally, electrical fires in the older building stock of all occupancy types, where wiring is often deteriorated and under-capacity for the modern loads applied, account for anywhere from 15 to 25 percent of all fires that occur in all buildings. Electrical fires are also the single most frequent cause of large-loss fires in the U.S. Large-loss fires are those resulting in over $1 million in property damage.

A second reason for the relatively severe fire statistics in this country has been ascribed to the lack of public awareness, concern, and ability to cope with fires in their early stages. It is sometimes said that this lack of concern, with regard to property losses and the handling of fire and its sources of ignition, is indigenous to people in the United States.

Mitigation of the first contributing cause requires that electrical systems be updated in the millions of buildings in this country that were constructed during the first half of this century. For new construction, it is highly desirable that premise wiring technology, especially branch wiring, which is responsible for most electrical fires, be advanced to incorporate monitors and fail-safe devices that can detect and prevent overloading, short circuits, and arcing phe-nomena, which are the three principal causes of electrical fires.

6.5.1 Minimum Requirements

In considering the role of building codes in connection with fire safety, it is important to recognize that these documents specify only minimum requirements, and that hazards associated with buildings of unusual design and housing diverse and multiple users create demands beyond such minimum requirements.

Modern designs that feature galleries, malls, atriums, and office and residential occupancies, all in a single building, vigorously test the sufficiency of design practices that simply comply with stated provisions of building codes.

Building codes are written, of necessity, in simple language, and generally do not treat the manner or means by which smoke control measures should be designed. Nor do they define, in sufficient detail, what a fully sprinklered building is or the possible consequences of exempting certain areas from sprinkler coverage.

For example, in the protection of atriums, the height of the ceiling or roof structure above the floor often precludes the installation of sprinkler heads at these otherwise normal locations. Because of this constraint, atrium floors often do not receive the benefit of sprinkler protection. Yet the building will still be deemed to be "fully sprinklered." These unprotected floor areas can be quite extensive in area, and may also contain extremely high and variable fuel loads, depending upon the use to which the space will be put at any given time.

It is also of relevance to note that the modern building code is usually a "performance" type code. This kind of code, rather the prescribing how a building is to be constructed, merely states performance requirements of building elements and components, and leaves it to the designer to select materials and methods of design and construction to meet that performance level.

This way of writing codes provides flexibility to the designer. Ostensibly, he can select the best materials and employ the most appropriate means for design for each project, without being limited by strict specifications that may not be appropriate under all conditions. On the other hand, performance codes leave much more to the judgment of the design professional and place a much greater burden on the inspection and certification process. Code enforcement also becomes more susceptible to breakdown, as it becomes more difficult for inspectors to identify clear violations.

The three uniform codes in use in the United States include the Uniform Building Code of the International Conference of Building Officials, The Standard Building Code of the Southern Building Code Congress, and the Basic National Building Code of the Building Officials Conference of America. It is said that these codes are the basis for 90 percent of all codes in force in the United States. In addition, many of the larger cities, which promulgate their own codes and regulations, use nationally recognized standards as references.

6.5.2 Standards

The Federal Trade Commission gives the following definition for a standard: ". . . . a technical document which describes design, material, processing, safety and performance characteristics of products." This definition suggests the far-ranging influence of standards on the ultimate safety of buildings from fire.

There are reported to be over 20,000 voluntary engineering standards published by approximately 360 organizations in the United States. This does not include military, government, and individual company standards. Altogether, it is estimated that there are over one million published standards in this country.

The multiplicity of both codes and reference standards in the United States leads to some difficulties as one views the rather substantial differences in requirements for buildings that are close to each other, but in different code jurisdictions.

There are no National Fire Protection Codes as such. The Federal government does issue standards that give minimum construction requirements for government properties and for projects that it entirely or partially sponsors and funds. Thus, there are FHA and HUD standards, as well as standards for hospital and other institutional buildings, which must be complied with.

For the greater part of building construction in the United States, the business of code promulgation is left to the states and local municipalities. This accounts for the wide variation in building and fire safety requirements found throughout the country. The states and other code writing bodies rely heavily on the model codes and consensus codes and standards developed by the National Fire Protection Association and known as the National Fire Codes. Unless officially adopted by a duly authorized governmental body, these codes and standards carry no legal weight, except perhaps during fire litigation proceedings. While consensus codes appear to work well, in general, their provisions must necessarily be acceptable to a mixed voting constituency that includes engineers and architects, fire officials, manufacturers, representatives from the insurance community, materials suppliers, and many other groups having an interest in building construction. This causes codes to evolve as compromises rather than as absolute measures directed to the life safety of occupants.

6.5.3 A Holistic Approach to Fire Safety

Fire Safety in Buildings. The decision tree for Fire Safety Systems Analysis, developed under the auspices of NFPA, has done much to unify, or at least to provide a common basis for understanding, the full breadth of fire safety and fire protection. The major subsystems that comprise fire safety are represented under a number of domains, all of which are interrelated in a comprehensive hierarchy flow diagram. The principal branches of the tree flow to such objectives as preventing fire ignition, or managing fire impact.

In turn, managing the impact of the fire can be addressed by managing the fire or managing the exposed. Figure 6.2 presents the full decision tree, as first presented in 1974.

6.6 EFFECTS OF FIRE ON BUILDINGS

6.6.1 Casualties from structural Failure

The factors of construction that lead to casualties in building fires generally do not involve a failure of structural materials or loss of structural integrity per se. Principal building framing, exterior walls, floors, and roofs that make up the basic structure and come under the responsibilities of the structural engineer generally fare well under fire conditions. This is because, with respect to the safety of occupants, these building elements need to remain functional and in place only for the relatively short period of time during which evacuation takes place. This is typically in the order of from 15 to 30 minutes.

Exceptions to the above may be associated with the use of wood trusses, whose failure under fire attack may cause massive collapse of roof structures. This hazard is especially critical to the safety of fire fighters, who often operate on roofs for some time after occupants have been safely evacuated. Such a failure occurred in 1978, when 100-foot long wood trusses installed within a double roof of a

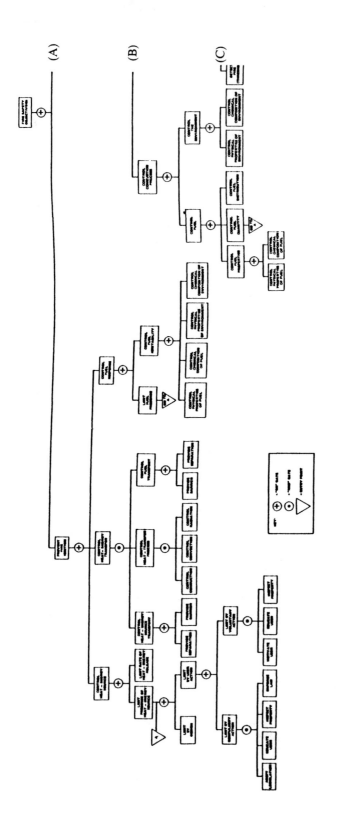

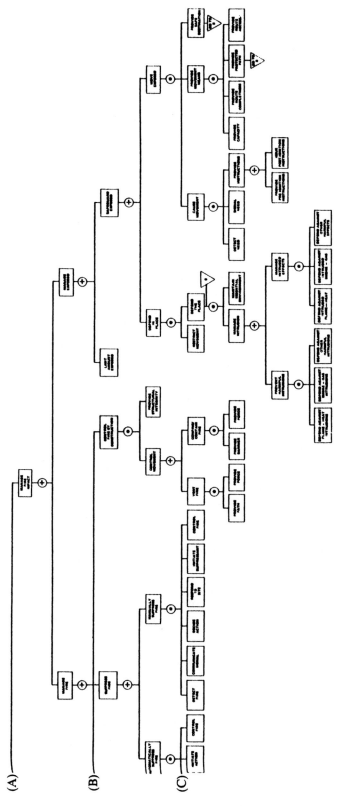

Figure 6.2. NFPA decision tree for fire safety systems analysis. Reprinted with permission from NFPA 550-86, *Guide to the Firesafety Concepts Tree*, Copyright© 1986, National Fire Protection Association, Quincy, MA 02269. This reprinted material is not the complete and official position of the National Fire Protection Association, on the referenced subject which is represented only by the standard in its entirety.

food supermarket in Brooklyn, New York collapsed during a fire, sending six fire fighters to their deaths.

The greatly expanded use, in recent years, of light-element wood trusses in residential construction, and the limited resistance that these constructions offer when used to support floors above cellars, are also of some concern when exposed to fires originating in the space below.

It is of some relevance to the issue of the safety of occupants to note that traditional factors of safety that were built into structures, when there was more doubt about the properties and uniformity of construction materials, and when analytical methods were not as powerful and reliable as they now are, are no longer present. Larger load-carrying members, more protective concrete and thicker slabs, of advantage during severe fires, have given way to lighter-weight constructions in all types of buildings.

While building collapse during fires tends not to be an important threat to occupants, construction details such as communicating concealed spaces, improperly lined shafts, inadequate fire-stopping, window and roof materials and details that favor the spread of fire, and a host of other items, can increase the rate of fire spread and the loss of normal escape routes.

Another concern, apart from building engineering, that greatly affects the likelihood of fire, as well as possibilities for its growth and extension, relates to the types, configurations, and placement of furnishings and other contents brought into the building. Here, wall and floor coverings, the presence and use of flammable liquids and combustible gases, and the placement, condition, and manner of the use of electrical appliances and wiring systems, all play prominent roles in fire scenarios resulting in the death and injury of occupants.

Nevertheless, while failure of the structure itself tends not to be a significant cause of casualties from fire, the design and operation of buildings with fire safety in mind can do much to offset other shortcomings and hazards that stem from a lack of regulation of contents, vagaries of human behavior, and modern lifestyles.

6.6.2 Structural Effects

As mentioned earlier, threats to life safety are not normally introduced through basic deficiencies of structural strength. By the time inherent construction weaknesses are manifested as a result of fire stresses in engineered structures, occupants have normally evacuated or have been rescued.

In some ways, the very fact that building structures are considerably overdesigned, with respect to occupant safety from fire, has introduced demands for a reduction of what is perceived by owners and some designers as overkill with significant impact on costs.

Current building codes give high assurance that buildings will not collapse or suffer other serious structural damage during the early stages of fires that are attended by modern fire departments. With the exception of wood frame residential buildings, the ability of interior walls and barriers to resist fires for periods of from 30 minutes to 1 hour is also reasonably certain in modern buildings.

The principal structural systems of a building that offer protection against fire are essentially the same as those that satisfy its basic qualities as a shelter. They include the basic framing elements and envelope, including the roof and walls. Interior walls, which may or may not be bearing walls, establish and protect corridors and enclose stairs and elevators. Interior fire-resistant walls also delimit compartments and prevent the spread of fire. Architectural elements and details that have important roles during fire incidents include windows, doors, shafts partitions, ceilings, and interior finishes.

Mechanical, electrical, and plumbing systems provide services necessary to operate the building. They include vertical transportation, lighting, power for all purposes, heating, ventilation and cooling, security and fire safety, and water supply and waste disposal. The performance of components and systems in each of these major design domains, as well as the manner of their installation, may greatly impact the safety of occupants during fires.

Barrier systems will delay the passage of flames, openings and penetrations are expected to be protected by water screens, noncombustible closures, or special types of glass. Shafts that penetrate floors are required to have walls of greater fire resistance, and barrier assemblies are required to be tested as a complete assembly, in approximately the same manner as it will be installed, and in accordance with approved and published protocols of the testing laboratory.

Floor and ceiling assemblies must be constructed to prevent vertical movement of flames from one floor to another, and openings made to facilitate construction and for the passage of conduits, wiring, ducts, and the like are required to be fire-stopped. When ducts penetrate compartment walls or floors and ceilings, smoke and fire dampers are usually prescribed, and concealed ceiling spaces are required to be free of combustible materials, with ceilings that will remain in place and retain their integrity under fire conditions for specified periods of time.

Columns, floors, partitions, walls, and other construction assemblies should be tested under both loaded and nonloaded conditions, and in restrained and unrestrained modes, as may be appropriate. The fire resistance of structural elements is substantially effected by the loading condition. For example, a concrete column stressed to 50 percent above design loads will lose approximately 55 percent of its fire resistance (68 minutes versus 124 minutes), and will have a fire resistance three-fold greater at 30 percent of its designed load than it will have under design load (358 minutes versus 124 minutes).

It is not uncommon to discover failure, during fires, of structural members that are subjected to undue loads generated by building extensions and alterations executed without the benefit of adequate engineering analysis.

Acceptance of structural components following fire testing is based upon the capability of test specimens to support design loads, their ability to limit temperature on unexposed surfaces (250°F), and to prevent the passage of heat or flame sufficient to ignite cotton waste or to exceed specified temperatures in steel members. Walls and partitions are normally also expected to maintain their integrity when exposed to a specified hose stream following fire testing.

It is well to recall that, in connection with the safety of occupants, required fire resistance of important structural elements usually ranges from one to four hours, depending on the construction classification and the use of the building under consideration, but that the behavior and performance of structural elements during the first 15 to 45 minutes or so following ignition (when evacuation, escape, and rescue operations are usually taking place) have the greatest significance, with respect to casualties. During this period, surface flame spread, flaking of walls and ceilings, the expulsion of hot particles from burning coatings, caulking, and sealants, undue levels and toxicity of smoke produced, the failure of construction details, including connections, fastenings, joints, and seams, doors and glass, while generally not classified as structural failure per se, are nevertheless critical to fire growth and the likelihood of escape.

Fire experience has shown that inadequate protection of steel and other structural materials, sometimes regarded as a detail of construction, has proven to be disastrous in fires. The performance under fire conditions of most structural materials is extremely sensitive to the thickness of the protective coverings and coatings.

It should also be remembered that test conditions ordinarily only approximate actual installations and fire situations, and that test specimens are carefully prepared, with workmanship of a quality not found at the construction site. The author has investigated a number of fatal fires in which the failure to properly fire-stop spaces adjoining flues and refuse tubes, and the failure to protect welding, burning, and brazing operations, have been responsible for the ignition of combustible materials through heat conduction.

6.6.3 Smoke Movement in Buildings

It is essential, at the outset of any consideration of the life safety of occupants, that building design professionals recognize that most fire fatalities and injuries result from smoke inhalation. Further, it must be acknowledged that each discipline involved in building design has certain responsibilities with respect to the mitigation of hazards associated with the development and migration of smoke during building fires.

For example, architects and structural engineers must be concerned with the locations and sizes of building compartments, and the number, location, and structural configuration of stair and elevator shafts, utility spaces, and vertical shafts. They must pay special attention to hidden spaces of all kinds, the size and locations of windows and other large air intake and exhaust openings, and the proximity of occupied spaces to large interior, and to domed roof, spaces. All of these features have direct and important influences on potential exposures of occupants to smoke and toxic gases.

The rate of smoke production in a fire may be estimated from expressions given in the *Fire Protection Handbook* (National Fire Protection Association, 1986) and elsewhere.

The mass of gas drawn into a fire may be expressed as:

$$M = 0.096 \, Pq_0 y^{\frac{3}{2}} \, (g \, T_0/T)^{\frac{1}{2}}$$

where

M = rate of smoke production (kg/s)
P = perimeter of the fire (m^3)
q_0 = density of ambient air (kg/m^3)
y = distance from floor to bottom of smoke layer (m)
g = acceleration due to gravity (m/s^2)
T_0 = absolute temperature of ambient air (°K)
T = absolute temperature of the flames in the plume (°K)

where the density of ambient air is taken as 1.22 kg/m^3 @ 17°C
absolute temperature of ambient air is 290°K;
absolute temperature of the flames of the smoke plume = 1100 °K;
gravitational constant = 9.81 m/s^2
    The expression reduces to:

$$M = 0.188 \, P \, y^{\frac{3}{2}}$$

The mass of smoke produced can be converted to volumetric measure by applying the appropriate factors, or by referring to tables provided in the above-cited publication. Since only the perimeter of the assumed fire, and the distance from the floor to the bottom of the smoke layer are required, the formula provides a first basis for dealing with the issue of smoke quantities to be addressed.
    Before considering factors that affect how smoke moves through buildings, and attempting to estimate or predict other quantitative aspects of the problem, it is necessary to give some thought to limits of smoke that can be tolerated.
    The questions of how much smoke and toxic gas components are too much, and how the design engineer may deal with the issue of permissible limits of smoke and gases, is difficult at best. There are, however, some useful reference points for developing a rational approach to these problems.
    For example, it is known that one of the unfortunate realities

relating to asphyxiation from carbon monoxide is that there is over a 200-fold greater affinity of hemoglobin for CO than there is between oxygen and this vital component of the blood.

It is also known that a level of carboxyhemoglobin (COHb) exceeding approximately 13 percent may be considered harmful. Some researchers place this level somewhat higher, at 20 to 25 percent. For the purposes of smoke control system design, the lower value appears to be more appropriate. This issue is of special importance to the protection of fire fighters, who must endure serial exposures to contaminated fire atmospheres within relatively short periods of time (De Cicco, 1982).

In order to limit COHb in the blood during occupant exposure to smoke and the gases of combustion, smoke exhaust and air supply rates must be controlled to limit CO accordingly. In this context, it is necessary to consider human exposures to relatively low levels of CO, in the order of several hundred ppm, for longer periods of time (30 minutes or more); as well as to exposures of much shorter duration (a few seconds to 30 minutes) at considerably higher concentrations (approximately 1 percent, or 10,000 ppm).

Concentrations of CO that may be expected in building atmospheres during fire conditions have been investigated by the Southwest Research Institute. Following a series of full-scale dwelling fire tests, they reported CO levels of 10,000 ppm (1 percent by volume) in from 7 minutes, 20 seconds to 25 minutes, 30 seconds following ignition.

In full-scale tests conducted by the author for the New York City Fire Department in a 22-story office building, CO levels in excess of 2000 ppm were measured in spaces in and adjacent to a 1700 square foot burn room, approximately 8 minutes after ignition (De Cicco et al., 1972). See Figure 6.3 for an illustration of the conditions under which this investigation took place.

More recently, in full-scale fire tests conducted in a number of row-frame residential buildings in Brooklyn, for the New York City Department of Housing Preservation and Development, the author observed levels of CO ranging from 1150 ppm to over 7000 ppm within the fire apartment, after periods from 2 minutes, 39 seconds to 11 minutes, 31 seconds following ignition. Since fuel types and loadings typical of those found in these high-risk neighborhoods

Figure 6.3. Full-scale fire tests in a high-rise office building ten seconds after ignition.

were used, and these are the very kinds of buildings in which most deaths and injuries from fire occur, the results are considered most relevant (De Cicco, 1976).

In two test fires conducted at the Henry Grady Hotel in Atlanta, Georgia, CO values measured by staff of the National Bureau of Standards reached levels of over 6000 ppm, and over 2000 ppm in rooms (with interconnecting duct work) adjacent to the burn room, in approximately 9 and 11 minutes following ignition. (Koplon, 1973). In a third test at the same location, approximately 2000 ppm of CO were measured in a room adjacent to the burn room 19 minutes after ignition.

In full-scale tests conducted at the Fire Research Station of the Building Research Establishment in the U.K., where the smoke release from a number of different types of wall linings was investigated, it was found that, in the case of all of the other materials examined, except for chipboard (fire insulating board, hardboard, polystyrene, plasterboard, glass reinforced polyester), only 5 to 10 minutes were required for the fire compartment and serving corridor to become smoke-logged (Wooley et al., 1978).

Tests conducted by the Fire Research Center of the National Bureau of Standards also contribute useful information concerning smoke leakage through suspended ceilings. These tests, conducted in a simulated hospital environment, used CO and smoke obscuration measurements to determine the hazards of smoke penetrating ceiling spaces and the means for reducing such risks (Klote, 1982).

Findings from both low-energy and high-energy tests indicate that downward flow into occupied spaces from above ceiling spaces can be prevented by exhausting such spaces at the rate of two air changes per hour. It is also of interest to note that differential pressures across ceilings were reported as approximately .036 inches of water (prior to the failure of the integrity of the ceiling suspension system). Smoldering fires produced no hazardous conditions in the concealed space.

These studies based permissible conditions on the relationship between CO concentrations in the environment and the level of COHb in the blood, in accordance with work by Stewart, as follows:

$$\text{COHb percent} = 5.98 \times 10^{-4} t \, [CO]^{1.036}$$

where t is exposure time in minutes, and CO is concentration of carbon monoxide in ppm.

This suggests that, at 2000 ppm, the COHb concentration would reach 20 percent in less than 13 minutes. As indicated earlier, the occurrence of CO levels in excess of 2000 ppm in rooms adjacent to the fire room is quite usual, and is often reached in the first few minutes after ignition.

During the above tests, it was also reported that, with the door to the burn room open and all ventilation systems inoperative, corridors and adjacent rooms were entirely obscured by smoke only two minutes after ignition. After approximately four minutes, the interstitial space rapidly filled with smoke.

The time to incapacitation (unconsciousness for a 70 kg male) as a result of exposure to CO has been estimated to vary from less than 2 minutes to almost 10 minutes in environments containing 1 percent CO, with the carboxyhemoglobin content ranging from 20 percent to 40 percent as activity levels during exposure vary from active to passive (Ross, 1984).

Measurements from full-scale tests demonstrate the relatively short period between the ignition of a fire and the time when levels of CO reach concentrations that pose serious threats to occupants. They also provide insights into the importance of eliminating or limiting dead-end corridor distances, providing the most direct and shortest travel distances to and through exits, and paying special attention to the possibilities of smoke travel through concealed spaces and ductwork.

In addition to occupant exposure to CO, failure to escape safely from fires may result from exposure to other products of combustion, including heat, to a lack of sufficient oxygen, and to the presence of other toxic gases and irritants.

Tenability with respect to temperature has been taken at three to five minutes at approximately 150 degrees Fahrenheit.

Times to incapacitation because of low oxygen are given as five minutes at 12 percent oxygen, and about the same time when CO, which increases respiration rate, reaches approximately 10 percent. Death will occur when oxygen falls to 7 percent.

It is also estimated that exposures to from 80 to 180 ppm of HCN, which is probably the most significant gas specie of building fires other than CO, will result in incapacitation in from 2 to 30 minutes. When concentrations exceed 180 ppm, less than 2 minutes are required to immobilize a person.

Although the smoke level, in terms of obscuration produced, does not directly incapacitate, it is nevertheless significant in that it causes disorientation and difficulties in locating stairs and exits. The extinguishment of a lightbeam over a given distance (1 m), known as the extinction coefficient, is the optical density per meter. It has been observed that people become disoriented when this value exceeds approximately 0.30. Measures of smoke density are also employed which depend upon the ability of a subject to read a standard sign at a given distance.

Eye and upper respiratory tract irritations resulting from exposure to smoke also impair escape capability, as these afflictions interfere with locating exits and tend to induce panic.

It can thus be seen that any structural or architectural feature (or omission thereof) that lengthens the time for exiting by only minutes, may contribute to casualties during fire emergencies. In making tenability and 'time available for escape' computations, it should also be understood that, in addition to the actual passage time or time in transit, there is often an appreciable time for initial occupant response and reaction to the first knowledge of the emergency.

It is obvious that large variations in materials and burning conditions in different fire situations give rise to wide range of rates of smoke and gas production. When the diversity of space configurations, smoke paths, and other environmental conditions are added, the dangers associated with over-generalization of either fire conditions or smoke control solutions become apparent.

REFERENCES

American Iron and Steel Institute. *Fire Protection Through Modern Building Codes*. New York: AISI. 1981.

De Cicco, P. R. *Life Safety Considerations in Atrium Buildings*. SFPE Technology Report 82-3. May 1982.

De Cicco, Paul R. What to do with existing row-frame residential buildings. *Fire Journal*, NFPA. November 1976.

De Cicco, P. R., Cresci, R. J., and Correale, W. H. *Pressurization and Exhaust in High-Rise Office Buildings*. Amityville, NY: Baywood Publishing Co. 1972.

Klote, J. *Smoke Movement Through a Suspended Ceiling System*. NBSIR 81-244, NBS Fire Research Center. February 1982.

Koplon, Norman A. Report of the Henry Grady Fire Tests, City of Atlanta, Georgia. January 1973.

National Fire Protection Association. *Fire Protection Handbook*, 16th Ed. 1986.

Ross, Steven S. *Construction Disasters*, Chapter 3. New York: McGraw-Hill Book Company. 1984.

Washa, G. W. *Concrete Construction Handbook*, Joseph J. Waddell, ed. New York: McGraw-Hill Book Company. 1974.

Wooley, W. D., Raftery, M. M., Ames, S. A., and Murrell, J. V. *Smoke Release from Wall Linings in Full-Scale Compartment Fires*.

Stewart, R. D. The effect of carbon monoxide on man. J. Fire and Flammability/combustion Technology. Vol. 1, 1974.

Chapter 7
Fire Hazard: Reducing the Consequences

Paul R. De Cicco, P.E.
Polytechnic University

The safety of occupants from building fires can probably best be enhanced through acknowledgement of obvious and time tested tenets long known to the fire community. In this context, there are a number of design features that should "come with the building" so as to bring simple, positive, and permanent qualities of safety to the structure and to those who occupy it.

Failure to provide such "inherent" safety features, derived from hard-learned rules, can only be compensated by introducing other safety measures that may not be required by codes and that are often overly articulated and complex. In any event, these responses and substitutions should be carefully planned and designed so that they will clearly, and on a one-for-one basis, provide safety levels equivalent to measures being replaced.

Unfortunately, methodology and rules for establishing such equivalencies are not well established, and to a large extent these kinds of trade-offs are much debated among fire officials, building officials, and applicants, when they are proposed.

With regard to the life safety of occupants, it should also be noted that most building codes include flexibilities and freedom for interpretation, which permits substantial digression from stated requirements of exit details, size of compartments, early warning systems, and sprinklerization. This situation warrants extreme caution on the parts of designers to avoid excesses in "trade-offs," even if regulatory agencies are willing to grant them.

Some examples of equivalencies or trade-offs permitted in various code jurisdictions include: reduced wall resistance in sprinklered buildings; omission of smoke control, such as pressurized fire stairs, in sprinklered buildings; larger compartments, when increased numbers of smoke detectors are used; reduced separation distances between buildings, where at least one building is protected by sprinklers; longer dead end corridors and longer distances to fire stairs.

7.1 BUILDING CONFIGURATION

Configurations of buildings are generally determined by architectural style, the need to provide a specified number of square feet of space on a given site, and the need to comply with zoning regulations and requirements for access and exiting.

The building shape and siting determine a structure's aerodynamic attributes, and therefore can significantly influence the movement of flames, heat, and smoke during fire incidents. Building setbacks can be important in providing accessible safe areas outside of the building, and can also provide additional opportunities for fire fighters to gain entry in rescue actions and to mount exterior attacks. On the negative side, setbacks can result in set-back stories becoming unreachable by ladder apparatus.

While it may be too much to expect that building configurations will be designed to optimize fire safety of occupants, designs should at least be evaluated for both the difficulties and opportunities they present with regard to life safety, and such information should be included in fire service preplanning, as well as in the development of the building's fire safety plan.

It should also be recognized that current tall building technology permits buildings to be constructed to almost any desired height, and structures 2000 feet tall are presently being contemplated. Aside from other aesthetic, sociological, and economic considerations that extremely tall buildings may engender, and in addition to the many well-recognized difficulties associated with locating and controlling high-rise fires, building height has two other major impacts on the fire safety of occupants.

The first relates to the strength of resulting stack action, which (if not controlled during a fire) can lead to extremely serious consequences with regard to smoke migration and its impact on occupants and fire fighters.

7.2 NEUTRAL PRESSURE PLANE

The nature and severity of stack action is determined largely by the location of the neutral pressure plane of a building. This horizontal reference plane passes through the intersection of a line representing the vertical variation of ambient pressures outside a building, and a line representing vertical variation of pressure within a building (often taken within stairs or other vertical shafts). At such an intersection, pressure differentials are neutral, and there is no tendency for air (or smoke) flow in either direction across the two domains.

Wherever the ambient pressure line represents pressures greater than those within the building (or shaft), flows will be into the building or shafts, and wherever the ambient pressure line represents pressures less than that within the building or shafts, flows will be outward from the building or shaft to the ambient environment.

The location of the neutral pressure plane is affected by the location and sizes of openings in the building, and there is some tendency to assume that it lies close to the vertical center of the structure. For a specific building, this can be far from its true position, and such oversight will make predictions of smoke movement directions quite unreliable.

The position of the neutral pressure plane of a building, when the upper opening is considered to be close to the top of the building, and the lower opening is close to the bottom, is given by:

$$h_1/h_2 = A_2 \, T_o/A_1 \, T_i$$

where:

h_1 and h_2 are the distances from the neutral pressure plane to the lower and upper openings;

A_1 and A_2 are the cross-sectional areas of the lower and upper openings; and

T_i and T_o represent the absolute temperatures of air inside and outside the building, respectively.

7.3 STACK EFFECTS

Stack action phenomena can produce the worst possible conditions with respect to smoke movement—driving smoke into stairs or shafts at locations below the neutral plane (during winter months when ambient air is cooler and denser than air within buildings), and forcing smoke out of stairs and other shafts and into occupied spaces at locations above the neutral pressure plane.

There are a number of alternative measures for smoke control that are designed to address problems associated with stack effects. They include stair pressurization, in which air is supplied to the stair to overcome the detrimental differential pressures described above; the use of stair vestibules, which act as locks to prevent smoke from entering fire stairs; and a number of other zone pressurization schemes, which attempt to produce positive pressure conditions on floors above and below the fire floor, and negative pressures on the fire floor (where mechanical exhaust will be provided).

In using any method for smoke control, building designers should be aware of complexities that attend most building fires and the difficulties involved in accounting for fire location, the locations of occupants and fire fighters, ambient pressures, the effects of sudden ventilation through windows, shafts, and through ceiling spaces, and pressures produced by the fire. Misapplication of mechanically induced or manually controlled gravity air movements can prove disastrous during building fires.

Stack action can make it impossible to open fire stair doors under certain conditions of ambient temperature and building system operation. Differential pressures in excess of one inch of water have been measured in winter time in lower lobbies of some of New York City's high-rise buildings. This results in over 100 pounds of pressure on a typical door and about half of this on the door knob.

The natural draft in a building may be estimated using:

$$D_t = 2.96 \ HB_o\rho \ (1/T_o - 1/T_i)$$

where:

D_t = theoretical draft, in inches of water
H = vertical distance between the inlet and the outlet, in feet
B_o = barometric pressure, in inches of mercury
T_o = temperature of outside air in degrees Rankine
T_i = temperature of inside air in degrees Rankine
ρ = density of air at 0°F and 1 atmosphere pressure in pounds per cubic foot

Using values of $B_o = 29.9$ in., and $\rho = 0.0862$ pcf, the expression reduces to:

$$D_t = 7.63H \ (1/T_o - 1/T_i)$$

7.4 WIND VELOCITY EFFECTS

A second major effect of building height stems from the increase in wind velocity at higher elevations. Wind velocities can vary as the 1/7 to 1/2 power of height. Wind pressures, in turn, vary with the square of velocity. A building 600 feet high might have wind velocities two and one-half times greater at the top floor than at the eighth floor level. At the same time, wind pressures might be six times greater at the top than at the eighth floor level.

In addition to exposing building interiors to extreme differential pressures (sometimes acting inwardly and sometimes causing external negative pressures, which drive fires out of window openings), when windows are broken by fire or by fire fighter actions, these higher winds effect the behavior of smoke and heat plumes emanating from the fire and can make helicopter and other roof rescue attempts difficult or impossible.

7.5 LARGE ENCLOSED SPACES

During the past two decades, there has been a large increase in the development of buildings that enclose extremely large volumes of contiguous space. These designs include multilevel shopping malls, galleries, covered sports arenas, convention halls, and atria, in all building occupancy categories.

The newly opened Marriott Marquis Hotel (1986), A&S Plaza Mall/atrium (1989), and the proposed Columbus Center multi-use projects in New York City have been planned with large spaces that enclose volumes of several million cubic feet. Figure 7.1, shows an elevator riding in an open atrium space.

7.6 ATRIUMS

The control of fire and smoke in buildings having large atrium spaces, which are interior covered courts penetrating one or more floors within the main structure, poses a number of special problems for fire protection engineers (De Cicco and Cresci, 1975). Inherently, these designs come into direct conflict with many recognized principles governing life safety in buildings. It has recently been estimated that one out of every four major buildings under construction in New York City contains one or more atrium spaces.

Figure 7.1. Glass-walled elevator within clear atrium space—New York Marriot Marquis Hotel.

While the atrium concept is simple, the ways in which these spaces are set within building structures are extremely varied and limited only by the imagination of the designer. Atriums are being constructed that range from single, centrally located spaces extending through the full height of the building, to multiple, stacked atria, in some cases interconnected, and in other instances entirely separated by intervening floors.

Multiple atria have also been designed that are distributed horizontally within the structure, that is, several independent spaces parallel to each other and extending vertically for several floors, or through the full height of the building. Configurations also vary from those in which the atrium is in full and open communication with adjacent occupied spaces, to those separated from other parts of the host structure by sprinkler-protected glass smoke barriers. Figure 7.2a shows stairs within the atrium at the NYU Bobst Library. In Figure 7.2b, glass separates the reading rooms from the atrium space at the NYU Bobst Library.

The physical scale and dimensions of modern atria also vary considerably, and (of great importance in the design of smoke control systems) these spaces have been built with volumes from less than 1000 cubic feet, to several millions of cubic feet. Floor areas, also of interest in the control of both fire and smoke, have ranged from only a few hundred square feet to tens of thousands of square feet. The heights of existing atrium constructions vary from less than 60 feet to over 600 feet.

Fire loads may also be uncertain in atriums and, given the wide range of uses of these spaces, it is important to anticipate potential loadings of combustible materials that may be quite different than first envisioned during the design stage.

With respect to internal geometry, atria have been built and are now under design which are simplistic in form, with enclosing walls that are relatively straight and smooth. Here, smoke flow patterns can be expected to be vertical and uninterrupted. In other instances, overhanging floors, large and sometimes complex architectural and structural forms, elevators, and ductwork may all act as deflectors, dead ends, and other obstructions to the flow of heat, smoke, and gases. As a result, smoke and gas concentrations may build up to levels considerably above those calculated or derived from generalized models. These same features may also significantly impact the effectiveness of smoke control systems. Figure 7.3 shows an atrium that is almost entirely filled with architectural constructions.

A number of serious fires in atrium buildings have been reported in the literature. Perhaps the most notable and revealing one occurred in the Regency Hyatt House Hotel in Rosemont, Chicago in 1973 (De Cicco and Cresci, 1975). In this fire, there was a clear demonstration of a number of problems associated with fire-protection systems in buildings with large central atria.

Exit signs were inadequate, alarms could not be heard within

Figure 7.2a. Stairs within atrium—Bobst Library, New York University.

Figure 7.2b. Glass separates reading areas from atrium in Bobst Library.

Figure 7.3. Westin Peachtree Plaza, Atlanta, GA—Architectural forms fill atrium space.

guest rooms, smoke exhaust fans were out of service, there were no smoke detectors within the room of fire origin, and every floor of guest rooms was open to the atrium space, which became smoke-logged.

Computations for the design and analysis of smoke control systems in atria have been developed over the past decade, which permit reasonable determinations of necessary exhaust rates for specific fire scenarios. Model studies have also been carried out in connection with the largest and most complex atria that have been built in the United States. Large-scale smoke tests have also been conducted prior to occupancy to evaluate the effectiveness of installed smoke control systems. Figure 7.4 shows a diagram of the New York Marriott Marquis Hotel, indicating smoke exhaust flows as tested in $\frac{1}{60}$-scale model.

The seriousness of failures of mechanical exhaust systems in atrium buildings demands that adequate and dependable standby power be provided, and that back-up gravity exhaust systems that can be manually operated be available.

It is important that gravity openings located at the top of the atrium, and sometimes at intervening points along its height, be of ample size, as the quantities of smoke to be handled may amount to hundreds of thousands of cubic feet per minute. At the same time, impelling buoyancies and pressure differentials may be small at interfaces between the upper atrium space and the exterior of the building, requiring substantial vent areas.

It should be especially noted that there can be no assurance that gravity venting systems alone will be sufficient to maintain safe conditions within an atrium or the adjacent communicating spaces. This is particularly true for smaller atrium volumes, in which dilution benefits are inconsequential.

In view of the criticality of smoke control in atriums, it is necessary to assure that, in mechanical venting systems, with certain motors and fans out of service for maintenance, remaining available capacities will be sufficient to provide the necessary rates of exhaust.

Some requirements and cautions regarding the fire protection of atriums may be summarized as follows:

1. Definition of dimensions and volumes of communicating spaces:

   Atriums are generally not limited in size or dimensions except that, in order to be designated an atrium, some codes may specify certain minimum floor areas and volumes; greater difficulties with respect to the time when conditions will become untenable, are likely to be encountered in smaller atriums.

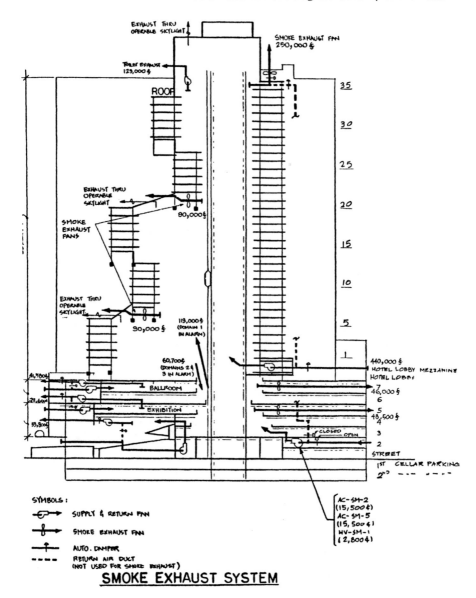

SMOKE EXHAUST SYSTEM

Figure 7.4. Large volume stacked atriums, model studies during design — Marriott Marquis Hotel, New York City.

Linear dimensions relating to potential smoke paths, the sizes of openings through which smoke may flow, and the furthest distances to exits, are all critical in atrium analysis.

A careful accounting of all spaces that communicate with the central atrium must be made, and modes of operation and the effects of mechanical air handling systems on spaces adjacent to the central atrium must be analyzed against appropriate fire scenarios.

2. Fire-suppression systems within and adjacent to atrium spaces:

Every attempt should be made to provide complete sprinklerization, including floor areas of the atrium per se; regulations against the use of sprinklers on atrium ceilings more than approximately 50 feet above the floor, in view of the extremely variable nature of potential fuel loads, should be addressed by installing high-pressure sidewall sprinklers or sprinklers on pedestals or other constructions available at the atrium floor.

3. Fire detection systems:

These should include automatic sprinkler water flow alarms and detectors placed at the atrium ceiling and at intervening points where smoke will pass or collect as it moves vertically through the atrium, and also in return air inlets and on the undersides of floors that project into the atrium space.

It is critical that the products of combustion be detected immediately, since even a relatively small fire can generate sufficient smoke and gases to log all but the largest of atrium volumes in less time than it takes fire services to arrive.

4. Fire alarms:

Including voice alarms, should be located in all spaces that communicate with the atrium. It is imperative that all occupants be able to hear such alarms, and careful consideration should be given to audibility in hotel rooms and other occupied spaces that may be threatened by fire or smoke in the atrium or by smoke moving towards exhaust points within the atrium.

5. Smoke control systems:

These must be capable of maintaining tenable conditions in all spaces from which or through which occupants will be passing on their way to a safe exit; exit times must be calculated for comparison with tenable times.

The locations and capacities of fan units that most codes require should be able to provide from 4 to 6 air changes per hour (depending upon volume of the atrium), and the quantity (50 to 75 percent of exhaust air, usually specified) and location of makeup air supply units are also critical to the effective operation of smoke control. The ability to operate supply and exhaust systems within the atrium and in adjacent connected spaces at 100 percent capacity must also be provided.

Most atria tend to be of unique design, and while code requirements should be observed as minimum needs, it is of the utmost importance that physical conditions, dilution capabilities, and operational modes of smoke control elements be designed and tested in the context of each specific project environment.

6. Activation of the smoke control system:

Codes do not always specify the manner in which actuation of smoke control systems will be initiated and, therefore, the sequence and triggering devices to be used deserve the careful consideration of designers. Normally, water flow alarms or the alarming of one or more smoke detectors will cause fan operation and dampers to be set; in all instances, codes will also require that all units be capable of manual control.

In New York City, it is required that all air handling systems go to the "off" mode on detection of fire by any means, and that the desired modes for exhaust, supply, and recirculation air units be determined by fire officials after appraisal of the fire situation.

7. Testing

After complete installation of the entire smoke control system, including all sensors, controls, air handling units, and emergency power units, and prior to the issuance of a certificate of occupancy, testing is essential.

A comprehensive test schedule and protocol that is directed to the evaluation of the performance of the smoke control system in the context of what has been specified in the approved "Fire Protection Plan" should be developed and implemented by the fire protection engineer, and witnessed by all those having authority and interest in the life safety of occupants in the subject building.

An appropriate ongoing test procedure and schedule should also be developed to assure that all devices, controls, and equipment remain in proper operating order during the life of the building.

8. Emergency and standby power facilities:

Emergency and standby power should be made available to operate all sensors, controls, and other equipment that are part of the smoke control system. These devices and equipment should be tested at regular intervals to assure readiness to perform.

9. Atrium enclosures:

Provisions for the number of floors permitted to open into the atrium (usually two or three), requirements for wall enclosures, and for the use of sprinkler-protected glass barriers, are specified in codes and should be adhered to.

Where such provisions are to be altered, it is imperative that analyses be performed to assure the equivalency of substitute measures.

10. Exiting:

Requirements for stairs within atria, and for the use of the atrium as part of required exiting, are important concerns, which vary somewhat in different codes. Calculated tenability characteristics during fire incidents should be used to determine how exiting will be designed in context of atrium spaces.

11. Uses of space:

Atria are amenable to diverse uses, with consequent extreme variations in fuel loads. The nature of atrium spaces is also such that relatively large numbers of persons may be assembled and original uses may change drastically both in the short-term or over longer periods of time; normally a low or ordinary hazard classification is assigned to atria, and some codes permit any use in atria that are completely sprinklered, which in turn demands that fire protection engineers anticipate a wide range of fire scenarios.

12. Interior finishes and decorations:

Because of the diverse uses of atrium spaces, finishes and decorations may be relatively inert with respect to fire characteristics, consisting of little more than noncombustible floors, walls, and ceilings; in other instances, exhibitions, stage entertainment, real or simulated plantings, and restaurant and café facilities may bring substantial fire hazards in the way of furnishings and floor, wall and ceiling treatments and coverings, and will thus deserve careful consideration.

7.7 "SLIVER BUILDINGS"

At the other extremity of the population of buildings that enclose large volumes are what have come to be known as "sliver" buildings. Born out of the growing value of central city land, these buildings are either built to disproportionate heights on very small plots, with only one or two residential apartments on each floor, or they are developed through the major renovation and upward expansion of existing buildings.

These structures present special problems with regard to the fire safety of occupants, among which are difficulties in protecting stairs and of fitting a second means of egress. In addition, gaining access to fight a fire at a floor above the reach of ladders in such buildings can present fire services with only single routes of interior attack, which greatly reduce the chances for timely rescue and fire control. Many building codes have recently been tightened to preclude the worst of these configurations.

7.8 COMPARTMENTATION

The importance of compartmentation as a basic means for limiting the spread and size of fire which must be handled merits the most careful consideration of building designers and fire protection engineers. While the sprinklering of large, open areas has proven effective, certain types of fuels embodied in furnishings, wall finishes and contents can tax conventional sprinkler systems, and the possibility of sprinkler system failure is always present.

The large loss of life in the casino fires in the MGM Grand Hotel in Las Vegas, and in the Dupont Plaza Hotel in San Juan, Puerto Rico, noted earlier, are suggestive of the special hazards that must be considered in all large, uncompartmented spaces.

In addition to limiting the size of building compartments, there is much to recommend that the number of compartments provided and their configuration, with respect to exits and other means of egress, be addressed. A floor area divided into only two compartments comprising 70 percent and 30 percent of the total floor area may well result in a fire in the larger compartment that will seriously limit the chances for escape, as well as constrain fire fighter access opportunities.

A fire involving 15,000 square feet in an upper story of a high-rise building will seriously challenge any fire departments' ability to control it. In short, compartment sizes and configurations must be considered on both an absolute basis to insure that fires that fill the compartment can still be controlled, and on a relative basis, which considers how escape routes and fire fighter access may be impacted by fire in various sized and located compartments.

The development of safe areas or areas of refuge is also considerably enhanced by multiple compartments, as this arrangement more often permits safe areas to be separated from the fire compartment by at least one other compartment or space. The complete sprinklering of a building that is also reasonably compartmented offers the best of all worlds, and gives up little in building cost or aesthetics for an extremely large return in safety.

7.9   MEANS OF EGRESS

The combination of the early warning of occupants and safe means of exit are perhaps the two prime factors for assuring escape from fires in buildings. The *Life Safety Code* (1985), published by NFPA as Standard 101, and entitled "Code for Safety to Life from Fire in Buildings and Structures," describes in detail the requirements for the design of exits.

Although primarily the concern of architects and fire safety specialists, there are a number of aspects of exit design that involve the structural engineer. For example, the final configuration of floor layouts at building corners may be influenced by the placement of columns, and by shafts and chases used to accommodate utilities. Structural layouts may create dead ends and other complexities of corridor alignment that allow only one path of egress for certain occupied areas, or which make paths and access to fire stairs less direct and less obvious.

The use of scissors stairs, used in the interest of simplicity and economy, can introduce serious deficiencies into the basic framework of occupant safety. In this type of design, two stairs are placed within a single shaft, sometimes with a separating wall between them and sometimes with no such separation. In this arrangement, if smoke or heat enters either stairway, there is the possibility that both stairs will be lost as a means of escape. In many instances, scissors stairs also compromise the principle of remoteness. For these reasons, their use should be restricted, or credit, in terms of the number of required stair units that they represent, should be discounted.

Exit capacity is generally determined by one of two methods, the so-called flow method and the capacity method. In the former, rates of flow of occupants along exit paths are based upon empirical data observed during actual evacuations and during evacuation drills.

Many factors influence flow rates, including: the age of evacuees, with the very young and elderly exhibiting the greatest variations from the norm; the season of the year, which has a bearing on clothing worn and space required per person; space available for each person; the width of passageways; and the size and location of obstructions which may be present. For example, slight obstructions have little influence, but dividing rails between travel lanes have an appreciable slowing influence on flow rates. Flow rates are generally taken as 60 persons per 22-inch exit width per minute, through level passageways and doors.

In the capacity method, it is required that there be sufficient stairways to house all building occupants with no flow out of the stairway. Under this concept, it is obviously imperative that the stair be well enclosed, have no penetrations that might admit smoke and, preferably, be protected by some means of pressurization to ensure that smoke cannot enter through doors that will be in use during an evacuation.

It is also critical for building designers to recognize that the entire path of exit must be safe during the full time that it is needed, and that, while there may be what is considered to be a primary route for exiting during fire emergencies, secondary or alternate paths that are required must offer equal safety in all respects. Unfortunately, this is not always the case, and such "alternate" paths are frequently inferior, in terms of accessibility, directness of route, obstructions encountered, and levels at which they are maintained.

The author has seen many examples of multiple, but certainly not equal, exiting facilities in the newest and largest of public buildings. They are the "back ways out," passing through service areas and "back-of-the-house" spaces that are unknown to occupants. Even assuming that sufficient exiting has been provided with respect to numbers, locations in the building plan and remoteness, there is still a critical need to be certain that exits are well marked, that doors are unlocked, at least during fire emergencies, and that occupants know where they are and can get to them under fire conditions.

In a hotel fire investigated by the author, exit signs were placed only over a single exit door located in a circular corridor. They could be seen only by those in a limited circular segment of the corridor. Instead of using the nearest and most appropriate stair, occupants travelled significant additional distances in attempts to retrace their steps on entering the hotel. To make matters worse, exit doors were painted to blend with surrounding wall colors, and most occupants passed these nearest and safest exits while searching for a way out.

In attempts to assure that occupants will be able to reach safe exits, new technologies have been offered that feature powerful strobe lights incorporated within exit signs, and some exits have been equipped with "talking signs," which beckon evacuees and also advise them of whether the stair is clear of smoke.

Simple and direct escape paths free of obstructions, careful consideration of special exit needs associated with each different occupancy served, and reference to data available in the technical literature concerning pedestrian flows in passageways of varying widths, lengths, and slopes will do much to optimize occupant safety.

Simple adherence to the letter of building codes and their mechanistic rules may often leave much to be desired with respect to occupant safety. This is amply demonstrated by a critical analysis of fire deaths and injuries that occur within exit paths in buildings in full compliance with exit code requirements.

7.10   OTHER MEANS FOR ESCAPE

In addition to more traditional forms of exits, a number of innovative means of escape from building fires have been offered over the years. Many of these have been developed in response to special hazards associated with fires in high-rise buildings. Some proposed methods and devices feature chutes or "stocking"-like flexible tubes, which are released from the point of escape and carry evacuees to safety at ground level.

Other systems utilize platforms that are affixed to the outside of the building and winched up or down, to the point of evacuation.

Still another escape system, developed in the Federal Republic of Germany, uses a truck-carried platform that is lifted to the desired rescue point by power from the truck that transports it. Cables attached to preset anchors on the building are used to move and elevate the unit into place.

Also developed in West Germany is a survival suit containing a 20-minute air supply. It is proposed that one or more units be placed in every hotel room to permit occupants to either reach a safe exit or remain in their room until rescued.

The McDonnell-Douglas Astronautics Company, using high technology, developed a Suspended Maneuvering System (SMS), commonly referred to as a "flying fire engine." This consists of a platform transported by helicopter to the fire site. The roof of the platform supports propulsion engines that can move the unit in any direction while it is being supported by a tether cable from the helicopter. The platform can be operated from either its own control station or from the helicopter. The SMS is capable of dispatching fire fighters to a fire site, or rescuing occupants from roofs, windows, or building ledges. The unit also carries extinguishants, hose lines, a gangway for transferring fire fighters or evacuees, and a deck nozzle.

While any of these devices would, indeed, be welcomed by an occupant stranded in a room or on a roof or ledge, in a sense these devices pay tribute to the failure of building designers to produce truly fire-safe structures. These types of devices do, however, offer possibilities for last-chance escape from existing buildings, which have exiting deficiencies that are not likely to be substantially improved because of attendant cost factors.

7.11 HUMAN BEHAVIOR

A substantial body of literature concerning human behavior during fires has been developed over the past decade or so. The movement of occupants in various building occupancies has been studied and observed during simulated evacuations, as well as during actual fire incidents. Most, if not all, of the important factors that affect safe exiting from buildings have been identified and verified through investigations, which have included on-scene interviews of evacuees and well-designed post-fire questionnaires (Canter, 1980; Kagawa et al., 1985; National Bureau of Standards, 1983; *Stahl, F. I.*, 1982).

Studies in the United States and elsewhere indicate that regular users of a building are more likely to attempt to extinguish a fire, alert others, and assist in evacuation than are persons with less familiarity, who tend to evacuate immediately.

Reports from Japan note that the amount of smoke observed is the principal determinant regarding escape routes chosen, but sex, position, and familiarity with the building also influence this decision. The latter factor is the most important with respect to speed and ease of evacuation.

7.12 FUTURE NEEDS

It is likely that trends in building design during the past two or three decades will continue in the future. This means that more innovative, larger, taller, and more complex structures, housing ever-greater numbers of people and containing multiple and diverse uses and activities, will continue to be designed and built.

It also means that there is little reason to expect an evolution of universal solutions to issues and problems regarding the life safety of occupants from fire. Finishes, decorations, and contents brought into buildings of all occupancy types can be expected to continue to vary widely in terms of their fire properties and the quantities of toxic smoke and gases that they will emit.

Technological advances in building design methodologies and in the use of structural materials to resist normal loadings are likely to bring further reductions in practices that favor redundancy in design and in the use of safety factors. These trends will further reduce contingency strengths and material masses of structural components and assemblies, and require that greater attention be focussed on the prevention of fires and on systems for early detection and the rapid suppression of fires that do occur.

The introduction and evolution of "smart buildings" in all occupancy categories will greatly increase the presence and significance of electrical and electronic systems. This thrust will bring new opportunities for the improvement of emergency communications, and for better management of fire safety and fire protection operations. However, at the same time, "smart buildings," as a result of proliferation of wiring, controls, and overall increase in the use of electrical energy, may be expected to bring new hazards to those already associated with contemporary electrical systems in buildings.

Perhaps another aspect of the emergence of "smart buildings" that bears watching is the inherent thrust of this technology towards automating everything in sight. In the case of fire phenomena in buildings, there is much need for a careful evaluation of each fire situation before altering environmental conditions that affect air supply, exhaust, recirculation, the operation of fire and smoke closures, and a host of other factors that can strongly influence the growth and manner of extension of fire and smoke migration in buildings.

In a recent National Research Strategy Conference sponsored by the National Fire Protection Association and the National Bureau of Standards, the full breadth of fire safety in the United States was examined by a large group of experts drawn from various sectors of the fire community. Participants represented the design professions, fire services, insurance industry, fire researchers from both public and private organizations, and materials suppliers and equipment manufacturers. Their purpose was to review and evaluate the state of the art of fire safety technology, and to recommend subject areas and strategies that should be addressed through national fire research efforts over the next ten years.

The recommendations that came out of this conference suggest how much still remains to be done in the interest of enhancing the life safety of occupants in buildings. Among the needs cited by the various sub-workshops were:

Design and Engineering—Improvements in fire modelling; making full-scale assessments of hazards from small-scale testing; fire risk assessment; the development of knowledge-based, performance-oriented codes and standards.

Fire Protection Systems—The systematic investigation of the effectiveness of active fire protection systems in all types of fires.

Building Technology—A critical assessment of technologies now in place to determine the performance and cost-effectiveness of present criteria; the development of models to predict fire assault on structures and structural response to fire.

Information Transfer—The development of engineering data bases and improved transfer mechanisms.

Materials and Products—Research into the fundamental properties of materials; material fire hazard assessment; a correlation of standard fire tests with performance in actual fires; solid phase degradation chemistry; soot and particulates; heat transfer from flames; the fire performance of building assemblies and extinguishment systems; the development of engineering methodologies for fire risk assessment of individual products or materials; the development of engineering methodologies for fire risk assessment of various systems, assemblies, and configurations of products; an evaluation of human behavior relative to materials and product use; an evaluation of the capability and reliability of products that reduce the risk of fire; the development of fire tests and models that accurately measure and predict the fire performance of products.

Howard Emmons, in a paper on "The Needed Science" presented at the first international symposium on Fire Safety Science (1985), also outlined work remaining to be accomplished. Some of the needs that he identified, relevant to the safety of occupants in buildings, are:

A better understanding of the ignition process, using additional parameters other than just temperature alone, and further consideration of layer and mixing theories

Enhanced general theories of radiation, flame spread, and charring phenomena that can be used with confidence when considering building fires

An investigation of the melting/burning characteristics of materials and the development of knowledge concerning soot production

A better understanding of the toxicity of pyrolysis gases and the nature of smoldering fires

An examination of the dynamics of flows to vents and through holes in ceilings and floors

A study of the behavior of window glass under fire conditions

Consideration of the action of fire brands in exposure fires, and an advancement of extinguishment science regarding the use of hose streams and sprinklers

7.13 SUMMARY

The determinants of life safety in specific buildings are numerous and complex. They relate to the structure, including the materials of construction, structural components, assemblies, and systems; to the utilization of space, including floor plans, type, number, and location of exits; to the combustibility of building finishes, furnishings, decorations, and contents; to the nature and circumstances of the threatening fire; and to the behavior under stress of occupants and the effectiveness of responding fire services.

In spite of the complexity of unwanted fire phenomena, over the past two decades research focussing on fire experience has established a substantial scientific base for an understanding of fire dynamics, has produced improved analytical tools for predicting the behavior of common building materials under fire conditions, and has provided bases for dealing with the problems of human exposure, behavior, and tenability in fire environments.

All that is needed in the way of technology, to reduce deaths and injuries from fires in buildings to a small fraction of those currently suffered, is already on hand (and has been for over a decade). What remains to be developed is a social determination that demands truly fire-safe buildings. The costs of sprinklerization, ample deployment of smoke detectors, generous exit facilities, and the control of hazardous materials that go into our buildings are small when compared with the tragic losses suffered by the nation each year.

REFERENCES

Backes, Nancy. *Great Fires of America*. Waukesha, WI Country Beautiful Corp. 1973.

David Canter, ed. *Building Evacuation, Research Methods and Studies, Fires in Human Behavior*. New York: John Wiley and Sons. 1980.

De Cicco, P. R. and Cresci, R. Smoke and fire control in large atrium spaces. *ASHRAE Proceedings*. 1975.

Kagawa, M., Kose, S., and Morishita, Y. Movement of people on stairs during fire evacuation drill—Japanese experience in a highrise office building. *Proceedings of the First International Symposium Fire Safety Science.* November, 1985.

National Bureau of Standards. *Implications for Codes and Behavior Models from the Analysis of Behavior Response Patterns in Fire Situations.* NBS-GCR-83-425. Washington, D.C.: National Bureau of Standards. March 1983.

National Fire Protection Association. *Life Safety Code,* NFPA 101. Quincy, MA: National Fire Protection Association. 1985.

Stahl, Fred I. BFIRES II, A behavior based computer simulation of emergency egress during fires. *Fire Technology.* Feb. 1982.

Chapter 8
Seismic Hazard

R.C. Murray
Lawrence Livermore National Laboratory
Livermore, California

and

M.K. Ravindra
EQE Engineering Inc.
Costa Mesa, California

8.1 INTRODUCTION

8.1.1 Categories of Hazards to Occupants

This chapter deals with the effects of earthquakes on structures and facilities, and the implications for the safety of their occupants. The hazards that the occupants are exposed to in an earthquake are mentioned in Chapter 1 of this book. Specifically, they can be categorized as follows:

- Failure of structural systems
- Failure of nonstructural systems
- Damage from unanchored equipment, furniture, etc.
- Loss of electricity and illumination

Well-engineered structures that are designed to modern building code requirements rarely experience gross structural failures threatening the safety of their occupants. This is borne out in a number of major earthquakes around the world. Structural failures in these earthquakes could, in almost all cases, be related to some gross human error in design or construction. Therefore, we could conclude that the probability of hazard to occupants of buildings in earthquakes from structural system failures to be insignificant.

The same cannot be said about the hazards from failures of nonstructural systems. Examples are failures of non-load-bearing walls, breakage of glass window panes, falling ceiling fixtures, and stuck-elevators. These events have occurred too often in earthquakes and have contributed the most to human fatality, injury, discomfort, and to property damage.

Most damage to the contents of a building in an earthquake occurs because of unanchored equipment, and due to furniture items, such as shelves and cabinets, which may empty their contents or fall over and impact with occupants or things stored in the building.

The last category of damage is the loss of electricity and illumination, which would hinder egress from the building in an earthquake or would slow down rescue operations.

8.1.2 Actions to Minimize Hazards

Although the building codes have been developed to reduce the risk of loss of life and limb to building occupants in an earthquake, some responsibility for minimizing the hazards in an earthquake should be borne by the occupants themselves. These include a general awareness about earthquakes and the proper responses during them, which include taking shelter under the desks and tables, in hallways and corridors, not using the elevator, etc. There are certain housekeeping activities that would also minimize the hazards. These include anchoring essential equipment, bracing shelves and cabinets, latching cabinet doors such that the contents may not fall out of them, etc. For some practical suggestions for the occupants, the reader is referred to the book by P. I. Yanev (1974). It is also recommended that a survey of the building or facility be performed by experienced engineers to identify the potential earthquake hazards and suggest mitigation measures. This is a very cost-effective approach for the treatment of earthquake hazards to occupants.

As building owners and designers, there are certain actions that could be taken to minimize the earthquake hazards to building occupants. In addition to following the explicit requirements of building codes, the designers should follow the code philosophy of providing ductile connections, alternate paths of load resistance, and redundancy. For discussions of these concepts, the reader is referred to the EERI report (1986) and ATC document (1978).

8.2 LIVERMORE VALLEY EARTHQUAKE, JANUARY 1980

The best way to demonstrate the effects of earthquakes on the safety of occupants is by example. The example used is from a moderate earthquake that occurred on January 24, 1980 in California's Livermore Valley. This earthquake caused damage to the Lawrence Livermore National Laboratory (LLNL), a large industrial facility located about 20 km (12 mi) from the epicenter. The earthquake measured 5.5 on the Richter scale, and produced peak horizontal ground accelerations at the site that were estimated to be between 0.15 and 0.3g. The earthquake was part of a sequence, which included two sharp aftershocks of magnitudes 5.2 and 4.2 within 15 minutes of the initial event.

Damage resulted primarily from the January 24, 1980 earthquake itself, with estimates of its economic impact at approximately $15 million (1980 dollars), including $3 million in direct damage and $12 million in upgrading costs. Five facilities that house radioactive materials suffered essentially no structural damage, while conventional structures (those designed according to the current Uniform Building Code) were damaged. The differences in behavior reflect differences in design criteria.

In the following, the impact of the January 24, 1980 earthquake is described, including some background information on LLNL, a discussion of the pre-earthquake seismic safety philosophy, and a description of the impact of the earthquake, with emphasis on the effects of the earthquake on building occupants. Improvements in the design criteria, personnel safety, and emergency preparedness that resulted from the experience are also indicated.

8.3 FACILITY BACKGROUND AND SEISMICITY

LLNL is located in Livermore, California, a community of 52,000 at the eastern end of the Livermore-Amador Valley. The site is about 50 miles southeast of San Francisco, California. LLNL began operations in 1952, as a research center supporting nuclear weapons development. Today, the facility has a population of about 8500, and conducts research and development for programs of national importance, such as nuclear weapons and future energy sources research. The site is one square mile, and has a variety of office and laboratory facilities. It is essentially a small city, with 150 buildings and over 1000 trailers. It has independent utilities (gas, water, electricity, telephone, and computer conduits), and its police, fire, and medical departments are maintained on-site. LLNL is operated by the University of California for the U.S. Department of Energy.

The area is surrounded by many small earthquake faults, and is within 50 miles of the large earthquake faults to the west. These include the San Andreas fault, which produced the great 1906 earthquake and fire that destroyed much of San Francisco, and the closer Hayward and Calaveras faults. The region is a seismically active part of California.

California has experienced many earthquakes since 1906, and these have led to improvements in seismic design requirements. In particular, the 1971 San Fernando earthquake (southern California) had stimulated an on-site program that led to tie-down and anchorage of much of the programmatic and facility equipment. Most of this anchorage work had been completed at the time of the 1980 earthquake. The 1978 Santa Barbara earthquake, which affected the University of California facilities at the Santa Barbara campus, pointed out the important lessons of tie-down and housekeeping. After the Santa Barbara event, earthquake awareness at LLNL was improved and the importance of being able to exit from facilities after earthquakes was emphasized. An aerial view of the site is shown in Figure 8.1.

8.4   THE JANUARY-FEBRUARY 1980 EARTHQUAKES

The primary cause of damage at LLNL was from the main shock on Thursday, January 24, 1980, which occurred at 11:00 AM. People were at work in their offices and laboratories. This was an earthquake of Richter magnitude 5.5, with an epicenter 20 km (12 mi) northwest of the site, along the Greenville Fault. The main shock was followed by two large aftershocks of magnitudes 5.2 and 4.2, which occurred within 1-½ minutes of the main shock. Ground motion from these events was focused toward LLNL, with the majority of the motion occurring in the east-west direction.

A large aftershock of magnitude 5.8 occurred two days later, on Saturday, January 26, 1980, with an epicentral distance of 13 km. This shock seemed to focus its energy away from LLNL and caused

Figure 8.1. Overview of Lawrence Livermore National Laboratory (*Courtesy of LLNL*).

little additional damage. Two additional aftershocks, large enough to be felt, with magnitudes around 2.6, occurred on February 12 and 21, 1980. The epicenters of these earthquakes are shown in Figure 8.2. This figure also shows some of the local fault systems surrounding the site and its location in California.

There were no strong-motion accelerometers on-site at the time of the earthquake. Ground motion estimates were developed from instruments in operation in the surrounding area, as well as from empirical relationships based on magnitude and epicentral distance, and observations of damage on-site. Table 8.1 indicates the information available to estimate the on-site ground motion.

It was estimated that the January 24, 1980 earthquake produced lateral forces on the structures and equipment at LLNL of approximately 25 percent of their weight. Some areas undoubtedly experienced higher values, due to amplification of the ground motions. For comparison purposes, this is similar to the safe shutdown earth-

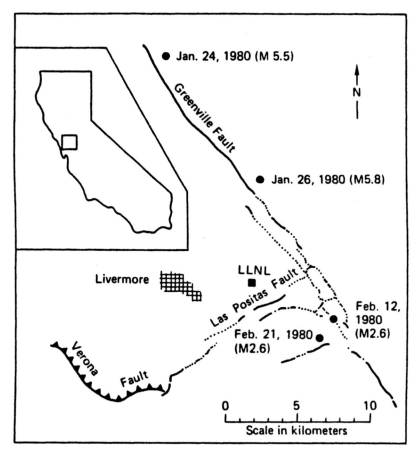

Figure 8.2. Faults in the Livermore Valley region (*Courtesy of LLNL*).

quake that is used to design nuclear power plants located in the eastern United States (taken as east of the Rocky Mountains). Nuclear facilities located in California have been designed for ground motions tabulated in Table 8.2.

A major difference between the January 24, 1980 earthquake at Livermore and the earthquake motion used for the design of U.S. nuclear power plants is the number of cycles of strong motion. The Livermore earthquake had only one strong spike, while design earthquakes are assumed to have many cycles of strong motion.

Table 8.1 Indicators, measurements in the surrounding area, and empirical data to estimate ground motion from the January 24, 1980 earthquake at LLNL.[1]

Observations		
Natural gas lines tripped		0.15g
Damage observed was caused by an earthquake with a modified		0.1–0.3g
Mercalli intensity of between VII and VIII		
Measurements in Surrounding Area (All were further than 20 km		
from the epicenter)		
Del Valle Dam	crest	0.26g
	toe	0.21g
VA Hospital		0.17g
GE Vallecitos		0.11g
Tracy		0.09g
Empirical Data		
Magnitude 5.6 earthquake epicentral distance of 13–16 km		0.15–0.25g

[1]Based on the available information, it was estimated that the ground motion produced on-site had horizontal accelerations of between 0.15g and 0.3g (*Note:* 1g is an acceleration equal to the acceleration of gravity, 980 cm/sec²).

8.5 SEISMIC DESIGN CRITERIA

The seismic design criteria used at LLNL has been revised many times during the past 40 years. The modifications reflect the impact of earthquakes and advances in knowledge and awareness concerning earthquakes. The history of the seismic design criteria used at LLNL is shown in Table 8.3. This criteria currently uses a site-specific response spectrum, which describes the frequency content of the design ground motion.

This criteria had been applied to the design of office and labora-

Table 8.2 Peak horizontal ground accelerations used to design nuclear facilities in California.

GE Vallecitos	0.75g
Diablo Canyon	0.75g
San Onofre Unit 2, 3	0.67g
Unit 1	0.50g
Rancho Seco	0.25g

Table 8.3 Increase in seismic design criteria used at LLNL during the last 40 years.

1940–1950	0.05–0.1 g	U.S. Navy preflight training station prior to LLNL occupying site. Uniform Building Code.
1950–1970	0.05–0.2 g	Uniform Building Code.
1970–1980	0.2g conventinal facilities 0.5 g critical facilities	Impact of the 1971 San Fernando earthquake in the southern California area.
1980	0.25 g design conventional facilities 0.50 g prevent collapse 0.50 g design critical facilities 0.80 g maintain integrity	Impact of the 1980 Livermore earthquake.

Figure 8.3. Overturned bookshelf in fallen typical office (*Courtesy of LLNL*).

Figure 8.4. Overturned bookshelf fallen in work area (*Courtesy of LLNL*).

tory facilities, and was used to anchor most of the large equipment used to conduct experiments and supply the plant engineering functions. Generally, the contents of the buildings do not meet any specific seismic criteria; they are placed where the occupants locate them based on functional requirements.

Prior to 1980, little attention was given to the seismic anchorage of office and laboratory furnishings such as bookcases and storage

Figure 8.5. Bookshelf fallen onto work area (*Courtesy of LLNL*).

racks, and to architectural details such as suspended ceilings and raised computer floors.

8.6   DAMAGE CAUSED BY EARTHQUAKE

The primary damage to LLNL was caused by the January 24, 1980 earthquake, and focussed on damage to furnishings in offices and laboratories and to architectural features. The most graphic illustration of this is shown in Figure 8.3. A typical office is shown in which a large book shelf has fallen on an occupant's desk. Fortunately, the occupant was not at the desk at the time of the earthquake. Notice the efficient storage scheme employed where the occupant had stacked bookshelves and cabinets on the desk as well as on top of other bookshelves. This is typical of many offices, where the occupants tend to have many references and store everything they have ever generated in their entire career. Housekeeping and some

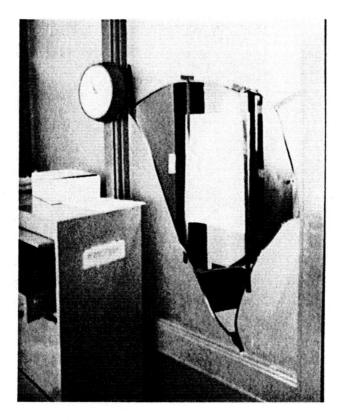

Figure 8.6. Broken glass in a partition wall (*Courtesy of LLNL*).

thought as to what could happen during an earthquake would obviously prevent this arrangement of office furnishings. Figures 8.4 and 8.5 show similar examples of overturning of large bookshelves, which were freestanding at the time of the earthquake. These examples further emphasize the need for good housekeeping, which is emphasized in earthquake awareness programs, and the need to protect oneself by taking cover under a desk or table or by moving to nearby doorways to avoid being injured by falling objects. Another example is broken glass used in interior windows or glass partition walls used in reception and clerical areas, as shown in Figure 8.6. Occupants should be aware of the possibility of injury due to large panes of broken glass, and arrange their office furniture to minimize injury and damage.

Figures 8.7 and 8.8 show damage that is beyond the control of the

Figure 8.7. Fallen suspended ceiling and light fixtures (*Courtesy of LLNL*).

Figure 8.8. A light fixture with heat return ducting shown fallen in a work area (*Courtesy of LLNL*).

Figure 8.9. Collapsed journal shelving in main library (*Courtesy of LLNL*).

Figure 8.10. Workmen assisting with rescue and clean-up operations (*Courtesy of LLNL*).

180

Figure 8.11. Collapsed bookshelves near library work area (*Courtesy of LLNL*).

Figure 8.12. Report storage shelving where contents were thrown into aisles (*Courtesy of LLNL*).

occupants and rests with the designers and contractors who build the facility. In these examples, suspended ceiling panels have fallen, and energy efficient lighting fixtures with heat return ducting have collapsed into office areas. Clearly, injury due to falling objects both within and beyond the occupants' control needs to be considered in ensuring earthquake safety.

One of the most spectacular examples of earthquake damage is shown in Figure 8.9, where the shelves containing the journals in the main library collapsed, throwing their contents into the aisles between the shelving. This situation required immediate attention to rescue people who may have been working in the area at the time of the earthquake and were possibly buried by books. Figure 8.10 shows workmen who were on the scene after the earthquake to assist with the rescue operations and cleanup. Fortunately, anyone in the stacks at the time of this earthquake escaped uninjured.

Figure 8.11 shows collapsed bookshelves near microfilm readers in the library. Figure 8.12 shows a report storage area with much of

Figure 8.13. Damaged journal shelves (*Courtesy of LLNL*).

the contents thrown into aisles, and finally, Figure 8.13 shows damage to other journal storage shelving. These examples indicate the need for adequate design and anchorage of shelving and the disruption that an earthquake can cause to normal operation. It was difficult to locate any particular journal for weeks after this earthquake. All of the books had to be removed, stored in temporary areas, the damaged shelving removed, and new shelving purchased and seismically braced before normal operations could resume.

Laboratory areas also have high injury and damage potential during earthquakes. Figure 8.14 shows a laboratory area where damage occurred to some of the sensitive instruments. Figure 8.15 shows a chemistry laboratory where most of the chemicals were thrown from their shelves and their containers broken. The clean-up crew in this area had to wear protective clothing with self-contained breathing gear, since some of the chemicals may have combined to produce toxic gases. Much of this damage could have been prevented by restricting the movement of chemicals on the shelving

Figure 8.14. Laboratory area where sensitive instruments were damaged by falling objects (*Courtesy of LLNL*).

Figure 8.15. Workmen in protective clothing cleaning up a chemistry laboratory.

either with sliding doors or other restraints. The use of plastic or unbreakable containers, where feasible, would also have prevented any chemical mixing.

Computer areas were another place where expensive damage could occur. Suspended ceiling panels can fall, and raised computer floors can collapse, if they are not adequately braced for lateral loads. Figure 8.16 shows a typical computer area. Overturning of unanchored computer cabinets can also occur. Designers of such facilities should pay special attention to anchorage and bracing details and to arrangements of components to provide stability against overturning.

8.7 SUMMARY

This chapter has described the different categories of hazards that the occupants of buildings are exposed to in earthquakes and how

Figure 8.16. Computer area where ceiling panels had fallen (*Courtesy of LLNL*).

they can take some measures to minimize the hazards. As an illustration, the effects of the January 1980 Livermore Valley earthquake on the LLNL facility are described, along with the measures that can be used to enhance occupant safety in future earthquakes.

REFERENCES AND SUGGESTED READING

Applied Technology Council. *Tentative Provisions for the Development of Seismic Regulations for Buildings.* ATC-3-06. June 1978.
ASCE Dynamic Analysis Committee. *The Effects of Earthquakes on Power and*

Industrial Facilities and Implications for Nuclear Power Plant Design. Prepared by Working Group on Past Behavior, 1987.

Eagling, D. G. *Seismic Safety Guide.* Lawrence Berkeley Laboratory report LBL-9143. September 1983.

Earthquake Engineering Research Institute. *Reducing Earthquake Hazards: Lessons Learned from Earthquakes.* Publication No. 86-02. November 1986.

EQE Incorporated. *Practical Equipment Seismic Upgrade and Strengthening Guidelines.* Lawrence Livermore National Laboratory report UCRL-15815. September 1986.

Murray, R. C., Nelson, T. A., Coats, D. W., Ng, D. S., and Weaver, H. J. Impact of the January-February 1980 earthquake sequence on various structures at the Lawrence Livermore National Laboratory, Paper K14/8. *Proceedings Sixth International Conference on Structural Mechanics in Reactor Technology*, Paris, France. August 1981.

The LLNL Earthquake Impact Analysis Committee Report on the Livermore California, Earthquakes of January 24 and 26, 1980. Lawrence Livermore National Laboratory report UCRL-52956. July 1980.

Yanev, P. I. *Peace of Mind in Earthquake Country: How to Save Your Home and Life.* San Francisco: Chronicle Books. 1974.

Appendix A
The Saffir-Simpson Hurricane Scale

Scale No. 1 — Winds of 74 to 95 miles per hour. Damage primarily to shrubbery, trees, foliage, and unanchored mobile homes. No real damage to other structures. Some damage to poorly constructed signs. And/or: storm surge 4 to 5 feet above normal. Low-lying coastal roads inundated, minor pier damage, some small craft in exposed anchorage torn from moorings.

Scale No. 2 — Winds of 96 to 110 miles per hour. Considerable damage to shrubbery and tree foliage; some trees blown down. Major damage to exposed mobile homes. Extensive damage to poorly constructed signs. Some damage to roofing materials of buildings; some window and door damage. No major damage to buildings. And/or: storm surge 6 to 8 feet above normal. Coastal roads and low-lying escape routes inland cut by rising water 2 to 4 hours before arrival of hurricane center. Considerable damage to piers. Marinas flooded. Small craft in unprotected anchorages torn from moorings. Evacuation of some shoreline residences and low-lying island areas required.

Scale No. 3 — Winds of 111 to 130 miles per hour. Foliage torn from trees; large trees blown down. Practically all poorly constructed signs blown down. Some damage to roofing materials of buildings; some window and door damage. Some structural damage to small buildings. Mobile homes destroyed. And/or: storm surge 9 to 12 feet above normal. Serious flooding at coast and many smaller structures near coast destroyed; larger structures near coast damaged by battering waves and floating debris. Low-lying escape routes inland cut by rising water 3 to 5 hours before hurricane center arrives. Flat terrain 5 feet or less above sea level flooded inland 8 miles or more. Evacuation of low-lying residences within several blocks of shoreline possibly required.

Scale No. 4 — Winds of 131 to 155 miles per hour. Shrubs and trees blown down; all signs down. Extensive damage to roofing materials, windows, and doors. Complete failure of roofs on many small residences. Complete destruction of mobile homes. And/or: storm surge 13 to 18 feet above normal. Flat terrain 10 feet or less above sea level flooded inland as far as 6 miles. Major damage to lower floors of structures near shore due to flood-

---

Reference: Herb Saffir Personal communication.

ing and battering by waves and floating debris. Low-lying escape routes inland cut by rising water 3 to 5 hours before hurricane center arrives. Major erosion of beaches. Massive evacuation of all residences within 500 yards of shore possibly required, and of single-story residences on low ground within 2 miles of shore.

Scale No. 5 — Winds greater than 155 miles per hour. Shrubs and trees blown down; considerable damage to roofs of buildings; all signs down. Very severe and extensive damage to windows and doors. Complete failure of roofs on many residences and industrial buildings. Extensive shattering of glass in windows and doors. Some complete building failures. Small buildings overturned or blown away. Complete destruction of mobile homes. And/or: storm surge greater than 18 feet above normal. Major damage to lower floors of all structures less than 15 feet above sea level within 500 yards of shore. Low-lying escape routes inland cut by rising water 3 to 5 hours before hurricane center arrives. Massive evacuation of residential areas on low ground within 5 to 10 miles of shore possibly required.

Appendix B
Estimation of Resistance Statistics

Let the random safety margin, Z, for a structural element be given by:

$$Z = R - S \qquad \text{B.1}$$

where, in general, R is the random resistance of the element and S is the random load. If we interpret S as a known design load, i.e., $S = R_{\text{design}}$, $\sigma_s = 0$, from Equation B.1, the reliability index, β, becomes:

$$\beta = (\overline{R} - R_{\text{design}})/\sigma_R \qquad \text{B.2}$$

where $\overline{R}$ is the mean value of the resistance. Since $\sigma_R = \overline{R}V_R$, where V_R is the coefficient of variation, Equation B.2 may be rewritten:

$$\overline{R}\beta V_R = \overline{R} - R_{\text{design}} \qquad \text{B.3}$$

Since β and V_R can be estimated from experience and calibration studies, we may solve for $\overline{R}$ in Equation B.3 to get:

$$\overline{R} = R_{\text{design}}/(1 - \beta V_R) \qquad \text{B.4}$$

Index

Printed in the United States
25022LVS00002B/299